Biodiversity of Vegetable Crops, A Living Heritage

Biodiversity of Vegetable Crops, A Living Heritage

Special Issue Editors

Massimiliano Renna
Pietro Santamaria
Angelo Signore
Francesco Fabiano Montesano
Maria Gonnella

MDPI • Basel • Beijing • Wuhan • Barcelona • Belgrade

MDPI

Special Issue Editors

Massimiliano Renna
Institute of Sciences of Food Production
CNR—National Research Council of Italy
Department of Agricultural and Environmental Science
University of Bari Aldo Moro
Italy

Pietro Santamaria
Department of Agricultural and
Environmental Science
University of Bari Aldo Moro
Italy

Angelo Signore
Department of Agricultural and Environmental Science
University of Bari Aldo Moro
Italy

Francesco Fabiano Montesano
Institute of Sciences of Food Production
CNR—National Research Council of Italy
Italy

Maria Gonnella
Institute of Sciences of Food Production
CNR—National Research Council of Italy
Italy

Editorial Office
MDPI
St. Alban-Anlage 66
4052 Basel, Switzerland

This is a reprint of articles from the Special Issue published online in the open access journal *Agriculture* (ISSN 2077-0472) from 2018 to 2019 (available at: https://www.mdpi.com/journal/agriculture/special_issues/biodiversity_vegetable_crops)

For citation purposes, cite each article independently as indicated on the article page online and as indicated below:

LastName, A.A.; LastName, B.B.; LastName, C.C. Article Title. *Journal Name* **Year**, *Article Number,* Page Range.

ISBN 978-3-03897-720-9 (Pbk)
ISBN 978-3-03897-721-6 (PDF)

Cover image courtesy of Beniamino Leoni.

Contents

About the Special Issue Editors

Massimiliano Renna is a researcher at the CNR-ISPA (Institute of Sciences of Food Production, National Research Council, Italy). In 2013 he received his Ph.D. degree in Mediterranean Agronomy at the University of Bari Aldo Moro, Italy. From 2015 to 2018 he was a researcher at the Department of Agricultural and Environmental Science, University of Bari Aldo Moro, Italy. He is an advisory board member of the Academy of Mediterranean Biodiversity Sciences and a member of the Italian Society of Horticulture. His research interests include a broad range of aspects regarding vegetable products, with particular attention being paid to vegetable landraces and wild edible plants. He is the author of three books and more than 60 scientific papers published in national and international journals and proceedings of national and international conferences.

Pietro Santamaria is an Associate Professor of the University of Bari Aldo Moro at the Department of Agricultural and Environmental Science. He teaches vegetable crops for two master's degree courses, soilless culture for the bachelor program "Agricultural Science and Technology", and science communication and scientific publications at the PhD School in "Biodiversity, Agriculture and the Environment". He coordinated the integrated project "Biodiversity of the vegetables of Puglia (BiodiverSO, 2014–2018)". He is the author or co-author of more than 300 scientific publications including fourteen books (six monographs).

Angelo Signore is a researcher of the Department of Agricultural and Environmental Sciences, University of Bari Aldo Moro (Italy). His research interests cover several aspects related to vegetables, including nitrates in leafy vegetables, soilless systems, agro-biodiversity, and the light spectrum. Together with Dr. Francesco Serio and Prof. Pietro Santamaria, he won the "Bram Steiner Award 2016" for outstanding research on the uptake of nutrients from nutrient solution by plants. He spent two years at Ghent University conducting studies on the influence of the light spectrum on the nitrate content in leafy vegetables. He is the author of more than 50 scientific papers published in national and international journals.

Francesco Fabiano Montesano is a researcher at the CNR–ISPA (Institute of Sciences of Food Production, National Research Council, Italy). After receiving his Ph.D. in Mediterranean Agronomy at the University of Bari 'Aldo Moro' (Italy), Dr Montesano has been focusing his research activities on the sustainable use of inputs and resources in horticultural processes. His areas of interest includes greenhouse cultivation, soilless techniques, sensor-based management of irrigation and fertilization, optimization of mineral plant nutrition for the enhancement of nutritional quality of vegetables, alternative growing media, and enhancement of the role of agro-biodiversity in food systems. He is involved in research projects and initiatives on the abovementioned topics at the national and international level, and plays an active role in the scientific community, both as an author and by regularly serving as a peer reviewer and supporting editorial processes for many international journals.

Maria Gonnella is researcher at the CNR-ISPA (Institute of Sciences of Food Production of the Italian National Research Council). Her research activity deals with vegetable production and quality under open field and greenhouse conditions, with particular attention paid to plant nutrition in relation to the organoleptic and nutritional quality of products and the sustainability of production. A wide range of species and varieties are the object of her studies, from wild genotypes to local landraces to cultivated varieties. She is the author of more than 40 scientific papers in national and international journals.

agriculture

MDPI

Editorial

Biodiversity of Vegetable Crops, A Living Heritage

Massimiliano Renna [1,*], Angelo Signore [2], Francesco F. Montesano [1], Maria Gonnella [1] and Pietro Santamaria [2,*]

[1] Institute of Sciences of Food Production, CNR – National Research Council of Italy, Via Amendola 122/O, 70126 Bari, Italy; francesco.montesano@ispa.cnr.it (F.F.M.); maria.gonnella@ispa.cnr.it (M.G.)
[2] Department of Agricultural and Environmental Science, University of Bari Aldo Moro, Via Amendola 165/A, 70126 Bari, Italy; angelo.signore@uniba.it
*   Correspondence: massimiliano.renna@ispa.cnr.it (M.R.); pietro.santamaria@uniba.it (P.S.); Tel.: +39-080-5443098 (P.S.)

Received: 27 February 2019; Accepted: 4 March 2019; Published: 6 March 2019

Abstract: Biodiversity is the natural heritage of the planet and is one of the key factors of sustainable development, due to its importance not only for the environmental aspects of sustainability but also for the social and economic ones. The purpose of this Special Issue is to publish high-quality research papers addressing recent progress and perspectives while focusing on different aspects related to the biodiversity of vegetable crops. Original, high-quality contributions that have not yet been published, or that are not currently under review by other journals, have been gathered. A broad range of aspects such as genetic, crop production, environments, customs and traditions were covered. All contributions are of significant relevance and could stimulate further research in this area.

Keywords: agriculture; landraces; local varieties; plant genetic resources populations; wild edible plants

1. Introduction

Intensive agriculture has generally determined a higher productivity, but also a decrease in agro-biodiversity, whose preservation represents a key-point to assure adaptability and resilience of agro-ecosystems to the global challenge (to produce more and better food in a sustainable way). Many components of agro-biodiversity would not survive without human interference, but human choices may also represent a threat for agro-biodiversity preservation [1].

The biodiversity in vegetable crops is underpinned by genetic diversity, which includes species diversity (interspecific diversity), the diversity of genes within a species (intraspecific diversity) that refers to the vegetable grown varieties, and by the diversity of agro-ecosystems (agro-biodiversity). Intraspecific diversity is very abundant in vegetable crops and is not reflected, at least not at the same extent, in other groups of crops. The labor operated by farmers over centuries of selection has led to the creation of a plurality of local varieties starting from the domestication of a wide agro-biodiversity forms, a precious heritage both from a genetic and a cultural-historical point of view. Therefore, the agro-biodiversity related to vegetable crops has assumed very articulated connotations. It is also important to specify that a "local variety" (also called a landrace, farmer's variety, or folk variety) is a population of a seed- or vegetative-propagated crop characterized by greater or lesser genetic variation, which is however well identifiable and which usually has a local name [1].

In facing the challenges of the modern vegetable growing sector, the many expressions of vegetable biodiversity are a key source for genetic improvement programs, and play a crucial role to produce innovative vegetables with improved qualitative characteristics (crop diversification and new crops), to realize more environmentally sustainable agro-systems, to cope with issues of climate change, and to find better adaptation to marginal soil conditions (salinity, atmospheric pollutants, etc.), while not forgetting the need to recover and maintain links with history and folk traditions [2].

Unfortunately, the genetic diversity of vegetable crops in many regions of the world has been eroded due to several factors such as abandonment of rural areas, ageing of the farming population, and failure to pass information down the generations (leading to loss of knowledge and historical memory), which can vary in relation to the type of genetic resource and location [1].

In this view, it is important to create a biodiversity network in order to promote the exchange of information between stakeholders and facilitate the diffusion and protection of these genetic resources by: collecting and preserving memories and knowledge of biodiversity in vegetable crops; retrieving and identifying such landraces wherever they can be found; characterizing, cataloguing and preserving them [3–5]. On the other hand, it must be underlined that the conservation of genetic biodiversity should be based not only on institutional and private plant breeders and seed banks, but mainly on the vast number of growers who continuously select, improve, and use vegetable biodiversity at the local scale. This availability of *in situ* biodiversity may be able to meet not only the requirements of breeders, but also the needs of specific niche markets, characterized by high demand for local products grown with environmentally-friendly farming techniques [3,4].

The purpose of this Special Issue is to publish high-quality research papers addressing recent progress and perspectives in different aspects related to the biodiversity of vegetable crops.

2. Papers in This Special Issue

The Special Issue "Biodiversity of Vegetable Crops, A Living Heritage" presents ten papers, focusing on a wide range of research activities and topics.

The first article concerns "A Protocol for Producing Virus-Free Artichoke Genetic Resources for Conservation, Breeding, and Production" by Roberta Spanò, Giovanna Bottalico, Addolorata Corrado, Antonia Campanale, Alessandra Di Franco and Tiziana Mascia [6]. This research article starts by suggesting that the potential of the globe artichoke biodiversity in the Mediterranean area is enormous but at risk of genetic erosion because only a limited number of varieties are vegetatively propagated and grown. In Puglia (Southern Italy), many globe artichoke ecotypes remained neglected and unnoticed for a long time and have been progressively eroded by several causes, including a poor phytosanitary status. This article describes a sanitation protocol based on the combination of in vitro meristem-tip culture and thermotherapy for the production and the *ex situ* conservation of virus-free propagation material of artichoke, in accordance with the EU Directives 93/61/CEE and 93/62/CEE. Five Apulian local varieties (Bianco di Taranto, Francesina, Locale di Mola, Verde di Putignano and Violetto di Putignano) were sanitized from artichoke Italian latent virus (AILV), artichoke latent virus (ArLV) and tomato infectious chlorosis virus (TICV) and a total of 25 virus-free primary sources were obtained and conserved *ex situ* in a nursery.

The second contribution explores "The Deterioration of Morocco's Vegetable Crop Genetic Diversity: An Analysis of the Souss-Massa Region" by Stuart Alan Walters, Rachid Bouharroud, Abdelaziz Mimouni and Ahmed Wifaya [7]. In this article, an assessment of the current status of vegetable landraces in the Anti-Atlas mountain areas of the Souss-Massa region (Southwestern Morocco) performed during 2014 is reported. It outlines how crop domestication and breeding efforts during the last half-century in developed countries significantly reduced the genetic diversity in Morocco. Results of this research showed that a significant loss of vegetable crop landraces (about 80-90%) has occurred in the last 30 years in this region of Morocco. Vegetable landraces that were notably lost during this time period included carrot (*Daucus carota* L.), fava beans (*Vicia faba* L.), melon (*Cucumis melo* L.), pea (*Pisum sativum* L.), watermelon (*Citrullus lanatus* L.), and especially tomato (*Solanum lycopersicon* L.). Authors highlighted that this genetic erosion will have a profound influence on future Moroccan agricultural productivity, since the genetic diversity within these landraces may be the only resource available to allow local farmers to cope with changing environmental conditions and optimize crop production in their harsh climate.

The third paper illustrates the "Quality and Nutritional Evaluation of Regina Tomato, a Traditional Long-Storage Landrace of Puglia (Southern Italy)" by Massimiliano Renna, Miriana Durante,

Maria Gonnella, Donato Buttaro, Massimiliano D'Imperio, Giovanni Mita and Francesco Serio [8]. Regina (Italian for "Queen") tomato, an Italian local variety grown in the coastal saline soils of the central Puglia, is listed as an item in the 'List of Traditional Agri-Food Products' of the Italian Department for Agriculture and itemized as 'Slow Food presidium' by the Slow Food Foundation. This local variety is classified as a long-storage tomato since it can be preserved for several months after harvest thanks to its thick and coriaceous skin. In this research article three ecotypes were investigated for the main physical and chemical traits both at harvest and after three months of storage. Experimental results indicate that Regina tomato has a qualitative profile characterized by high concentrations of tocopherols, lycopene and ascorbic acid even after a long storage time, as well as lower average glucose and fructose contents compared to other types of tomatoes. Authors highlighted the high nutritional value of this local variety, especially for people with specific dietary requirements, as well as the possibility to use these results as a tool for obtaining the Protected Geographical Indication or Protected Designation of Origin mark.

The fourth article is the "Cultivation of Potted Sea Fennel, an Emerging Mediterranean Halophyte, Using a Renewable Seaweed-Based Material as a Peat Substitute" by Francesco Fabiano Montesano, Concetta Eliana Gattullo, Angelo Parente, Roberto Terzano and Massimiliano Renna [9]. Sea fennel (*Crithmum maritimum* L.) is a halophyte species belonging to the *Apiaceae* family that is used in folk medicine as well as in many traditional dishes for its interesting sensory attributes. For this research, the authors used three posidonia (*Podisonia oceanica* (L.) Delile)-based composts (a municipal organic solid waste compost, a sewage sludge compost, and a green compost) by hypothesizing that the halophytic nature of sea fennel allows to overcome the limitations of high-salinity compost-based growing media. Results of this research article show the possibility of using posidonia compost-based substrates without any negative effect on the sea fennel growth in comparison with a commercial peat substrate. Therefore, the authors suggest that these substrates can be used as a sustainable peat substitute for the formulation of soilless mixtures to grow potted sea fennel plants, even up to a complete peat replacement.

The fifth contribution titled "Phytochemical Analysis and Antioxidant Properties in Colored Tiggiano Carrots" by Aurelia Scarano, Carmela Gerardi, Leone D'Amico, Rita Accogli and Angelo Santino" [10] assesses the content of carotenoids, anthocyanins, phenolic acids, sugars, organic acids, and antioxidant activity in a carrot landrace of Southern Italy also called Carrot of Saint Ippazio. The authors indicated that this yellow-purple carrot has a higher level of bioactive compounds, together with the highest antioxidant capacity, compared to the yellow and cultivated orange varieties. These results point out the nutritional value of Tiggiano carrots and may contribute to the valorization of this local variety. Moreover, the presence of bioactive compounds highlights on the possible activation of the anthocyanin biosynthetic pathway in the taproots.

The sixth article is "Conservation of Crop Genetic Resources in Italy with a Focus on Vegetables and a Case Study of a Neglected Race of Brassica Oleracea" by Karl Hammer, Vincenzo Montesano, Paolo Direnzo and Gaetano Laghetti [11]. In this study the authors provide a summary of the conservation strategies for autochthonous agrobiodiversity in Italy with a special focus on vegetables. The paper also offers an outlook on the most critical factors of *ex situ* conservation and actions which need to be taken. Some examples of 'novel' recovered neglected crops are also given. Finally, a case study is proposed on 'Mugnolicchio', a rare landrace of *Brassica oleracea* L., cultivated in Southern Italy, that might be considered as an early step in the evolution of broccoli (*B. oleracea* L. var. *italica* Plenck).

The seventh contribution concerns "Issues and Prospects for the Sustainable Use and Conservation of Cultivated Vegetable Diversity for More Nutrition-Sensitive Agriculture" by Gennifer Meldrum, Stefano Padulosi, Gaia Lochetti, Rose Robitaille and Stefano Diulgheroff [12]. This study reviewed the uses, growth forms and geographic origins of cultivated vegetables worldwide and the levels of research, *ex situ* conservation, and documentation they have received in order to identify gaps and priorities for supporting a more effective use of global vegetable diversity. A total of 1097 vegetables were identified in a review of the Mansfeld Encyclopedia of Agricultural and Horticultural Plants.

The article reports that documentation for most vegetable species is poor and the conservation of many vegetables is largely realized on farms through continued use. Therefore, the authors suggest that supportive policies are needed to advance research, conservation, and documentation of neglected vegetable species to protect and further their role in nutrition-sensitive agriculture.

The eighth paper illustrates "BiodiverSO: A Case Study of Integrated Project to Preserve the Biodiversity of Vegetable Crops in Puglia (Southern Italy)" by Massimiliano Renna, Francesco F. Montesano, Angelo Signore, Maria Gonnella and Pietro Santamaria [13]. Puglia region (southern Italy) is particularly rich in agro-biodiversity, representing an example of how local vegetable varieties can still strongly interact with modern horticulture [1]. This article summarizes the objectives, methodological approach and results of the project "Biodiversity of the Puglia's vegetable crops (BiodiverSO)", an integrated project funded by Puglia Region Administration under the 2007–2013 and 2014–2020 Rural Development Program. Moreover, a case study is proposed on the Polignano carrot, a multicolored landrace of *Daucus carota* L. strongly linked with local traditions. Overall, the authors suggest that *in situ* conservation of genetic resources needs to be based not only on institutional programs, but mainly on the possibility, especially for young growers, to use these resources for productive activities which would facilitate a real income.

The ninth paper regards "Patterns of Genetic Diversity and Implications for In Situ Conservation of Wild Celery (*Apium graveolens* L. ssp. *graveolens*)" by Lothar Frese, Maria Bönisch, Marion Nachtigall and Uta Schirmak [14]. This study has been carried out to support *in situ* conservation actions regarding the wild ancestor of celery and celeriac in Germany. Seventy-eight potentially suitable genetic reserve sites representing differing eco-geographic units were assessed with regard to the conservation status of the populations. The authors determined the structures of genetic diversity within the sampled material as well as the differences in trait distribution between occurrences. The article recommended that 15 sites among those identified be used to form a genetic reserve network. This organizational structure appears to be suitable for promoting the *in situ* conservation of intraspecific genetic diversity and the species' adaptability. Moreover, the authors highlight that genetic reserves are conservation projects that require the support and active collaboration of local people without which a genetic reserve can neither be established nor maintained over an effectively long period.

The tenth article concerns the "Diversity of Cropping Patterns and Factors Affecting Homegarden Cultivation in Kiboguwa on the Eastern Slopes of the Uluguru Mountains in Tanzania" by Yuko Yamane, Jagath Kularatne and Kasumi Ito [15]. This paper focuses on the diversity of the cropping pattern observed in homegardens distributed on the eastern slopes of the Uluguru Mountains (Central Tanzania) and how this diversity developed. Participatory observation with a one year stay in the study village was conducted to collect comprehensive information and to detect specific factors about formation of diversity cropping patterns of homegardens. The authors indicated that the diversity of the cropping patterns observed in the homegardens in the target study village was influenced by factors related to regional characteristics such as the regional history and the customs and policies. In addition, ecological diversity, distributed on the slopes of the mountains from around 650 m to around 1200 m, also makes the cropping pattern diverse. The authors also highlighted a useful inductive method of analysis to facilitate an essential understanding of this ecological diversity.

3. Conclusions

In conclusion, the papers of this special issue cover a broad range of aspects and represent some of the recent research results regarding the topic of agro-biodiversity, which continues to be significantly relevant for both genetic and agriculture applications. We think that this special issue may stimulate further research in this area.

Author Contributions: All coeditors equally contributed to organizing the special issue, to editorial work, and to writing this editorial.

Acknowledgments: We thank the authors for submitting manuscripts of high quality and their willingness to further improve them after peer review, the reviewers for their careful evaluations aimed at eliminating weaknesses and their suggestions to optimize the manuscripts and the editorial staff of MDPI for the professional support and the rapid actions taken when necessary throughout the editorial process.

Conflicts of Interest: The authors declare no conflict of interest.

References

1. Elia, A.; Santamaria, P. Biodiversity in vegetable crops, a heritage to save: The case of Puglia region. *Ital. J. Agron.* **2013**, *8*, 4. [CrossRef]
2. Signore, A.; Renna, M.; Santamaria, P. Agrobiodiversity of vegetable crops: Aspect, needs, and future perspectives. *Annu. Plant Rev.* **2019**, *2*, 1–24.
3. Renna, M. From the farm to the plate: Agro-biodiversity valorization as a tool for promoting a sustainable diet. *Prog. Nutr.* **2015**, *17*, 77–80.
4. Renna, M.; Serio, F.; Signore, A.; Santamaria, P. The yellow-purple Polignano carrot (*Daucus carota* L.): A multicoloured landrace from the Puglia region (Southern Italy) at risk of genetic erosion. *Genet. Resour. Crop Evol.* **2014**, *61*, 1611–1619. [CrossRef]
5. Testone, G.; Mele, G.; Di Giacomo, E.; Gonnella, M.; Renna, M.; Tenore, G.C.; Nicolodi, C.; Frugis, G.; Iannelli, M.A.; Arnesi, G.; et al. Insights into the sesquiterpenoid pathway by metabolic profiling and de novo transcriptome assembly of stem-chicory (*Cichorium intybus* Cultigroup "Catalogna"). *Front. Plant Sci.* **2016**, *7*, 1676. [CrossRef] [PubMed]
6. Spanò, R.; Bottalico, G.; Corrado, A.; Campanale, A.; Di Franco, A.; Mascia, T. A protocol for producing virus-free artichoke genetic resources for conservation, breeding, and production. *Agriculture* **2018**, *8*, 36. [CrossRef]
7. Walters, S.A.; Bouharroud, R.; Mimouni, A.; Wifaya, A. The deterioration of Morocco's vegetable crop genetic diversity: An analysis of the Souss-Massa Region. *Agriculture* **2018**, *8*, 49. [CrossRef]
8. Renna, M.; Durante, M.; Gonnella, M.; Buttaro, D.; D'Imperio, M.; Mita, G.; Serio, F. Quality and nutritional evaluation of regina tomato, a traditional long-storage landrace of Puglia (Southern Italy). *Agriculture* **2018**, *8*, 83. [CrossRef]
9. Montesano, F.F.; Gattullo, C.E.; Parente, A.; Terzano, R.; Renna, M. Cultivation of potted sea fennel, an emerging Mediterranean halophyte, using a renewable seaweed-based material as a peat substitute. *Agriculture* **2018**, *8*, 96. [CrossRef]
10. Scarano, A.; Gerardi, C.; D'Amico, L.; Accogli, R.; Santino, A. Phytochemical analysis and antioxidant properties in colored Tiggiano carrots. *Agriculture* **2018**, *8*, 102. [CrossRef]
11. Hammer, K.; Montesano, V.; Direnzo, P.; Laghetti, G. Conservation of crop genetic resources in Italy with a focus on vegetables and a case study of a neglected race of Brassica Oleracea. *Agriculture* **2018**, *8*, 105. [CrossRef]
12. Meldrum, G.; Padulosi, S.; Lochetti, G.; Robitaille, R.; Diulgheroff, S. Issues and prospects for the sustainable use and conservation of cultivated vegetable diversity for more nutrition-sensitive agriculture. *Agriculture* **2018**, *8*, 112. [CrossRef]
13. Renna, M.; Montesano, F.F.; Signore, A.; Gonnella, M.; Santamaria, P. BiodiverSO: A case study of integrated project to preserve the biodiversity of vegetable crops in Puglia (Southern Italy) *Agriculture* **2018**, *8*, 128. [CrossRef]
14. Frese, L.; Bönisch, M.; Nachtigall, M.; Schirmak, U. Patterns of genetic diversity and implications for in situ conservation of wild celery (*Apium graveolens* L. ssp. *graveolens*). *Agriculture* **2018**, *8*, 129. [CrossRef]
15. Yamane, Y.; Kularatne, J.; Ito, K. Diversity of cropping patterns and factors affecting homegarden cultivation in Kiboguwa on the Eastern Slopes of the Uluguru mountains in Tanzania. *Agriculture* **2018**, *8*, 141. [CrossRef]

agriculture

MDPI

Article

A Protocol for Producing Virus-Free Artichoke Genetic Resources for Conservation, Breeding, and Production

Roberta Spanò [1,2,*], Giovanna Bottalico [1], Addolorata Corrado [1], Antonia Campanale [2], Alessandra Di Franco [1] and Tiziana Mascia [1,2]

[1] Dipartimento di Scienze del Suolo della Pianta e degli Alimenti, Università degli Studi di Bari "Aldo Moro", Via Amendola 165/A, 70126 Bari, Italy; giovanna.bottalico@uniba.it (G.B.); corrado.ada82@gmail.com (A.C.); alessandrarosari.difranco@uniba.it (A.D.F.); tiziana.mascia@uniba.it (T.M.)
[2] Istituto per la Protezione Sostenibile delle Piante (IPSP), CNR, UOS Bari, Via Amendola 122/D, 70126 Bari, Italy; antonia.camapnale@ipsp.cnr.it
* Correspondence: roberta.spano@ipsp.cnr.it; Tel.: +39-080-544-3086

Received: 30 January 2018; Accepted: 27 February 2018; Published: 1 March 2018

Abstract: The potential of the globe artichoke biodiversity in the Mediterranean area is enormous but at risk of genetic erosion because only a limited number of varieties are vegetatively propagated and grown. In Apulia (southern Italy), the Regional Government launched specific actions to rescue and preserve biodiversity of woody and vegetable crops in the framework of the Rural Development Program. Many globe artichoke ecotypes have remained neglected and unnoticed for a long time and have been progressively eroded by several causes, which include a poor phytosanitary status. Sanitation of such ecotypes from infections of vascular fungi and viruses may be a solution for their ex situ conservation and multiplication in nursery plants in conformity to the current EU Directives 93/61/CEE and 93/62/CEE that enforce nursery productions of virus-free and true-to-type certified stocks. Five Apulian ecotypes, Bianco di Taranto, Francesina, Locale di Mola, Verde di Putignano and Violetto di Putignano, were sanitized from artichoke Italian latent virus (AILV), artichoke latent virus (ArLV) and tomato infectious chlorosis virus (TICV) by meristem-tip culture and in vitro thermotherapy through a limited number of subcultures to reduce the risk of "pastel variants" induction of and loss of earliness. A total of 25 virus-free primary sources were obtained and conserved ex situ in a nursery.

Keywords: artichoke; ecotype; virus-sanitation; meristem-tip culture; thermotherapy

1. Introduction

Globe artichoke (*Cynara cardunculus* L. var. *scolymus*) is a species native to the Mediterranean basin that is gaining commercial interest for its dietary and pharmaceutical value. Out of the Mediterranean area, production occurs also in the Middle East, North Africa, South America, United States, and China.

Italy is the leading world producer of globe artichoke and probably hosts the most abundant in situ diversity of the species. Apulia (southern Italy) (Figure 1) accounts for 33% of the total Italian production estimated at 470,000 tons (FAOSTAT 2011) and harbors many artichoke varieties representing an inestimable source of germplasm [1–5]. Most of the germplasm has remained neglected and unnoticed for a long time and has been progressively eroded by several causes [6], which include a poor phytosanitary status [7]. This condition stalled until the Apulian Regional Government launched two specific actions in the framework of the 2007–2013 Rural Development Program (RDP): action 214/3 "Protection of biodiversity" and action 214/4 "Integrated projects and regional biodiversity system". Both the actions were aimed at identifying, rescuing and preserving biodiversity of woody and vegetable crops grown in rural areas of the Region. Globe artichoke was included in the list as one of the vegetable species recognized at risk of genetic erosion.

Compared with the abundance and diversity of the ecotypes available, only a limited number of varieties are grown in Apulia, which is, by itself, indicative of genetic erosion. Most of the ecotypes probably belong to the same genetic background but their name derived from the place where they were traditionally cultivated.

Globe artichoke is grouped as early-flowering or late-flowering type based on the time the capitula appear. Early-flowering types cluster with the "early Mediterranean group" and produce capitula between autumn and early spring, if water is supplied to dormant underground shoots and buds during summer. Conversely, late-flowering types produce capitula during spring and early summer. Despite the recent introduction of new variety types propagated by seeds, vegetative propagation through shoots and buds remains the preferred method for the new plantings of traditional globe artichoke varieties.

Implementation of globe artichoke propagation material with unknown phytosanitary status has led to the transmission and accumulation in the same plant of threatening pathogens such as vascular fungi and viruses for which no efficient control measures are available [7]. Over time, severe production losses and reduced quality have led to the shifting of crops from the south of the Bari area to the north of the region, in the Foggia province (Figure 1). However, since the material used for the new plants was infected by the vascular fungus *Verticillium dahliae* and by symptomless viruses these pathogens were simply transferred to the new growing areas with transitory and poor benefits to the globe artichoke market. The virome (*sensu* [8]) of globe artichoke may be unexpectedly complex as it includes at least 25 virus species in 10 taxonomic families that adversely affect both quality and productivity of the crop (Table 1). Some of the viruses are widespread whereas others are endemic of the Mediterranean area and frequently asymptomatic in globe artichoke. Additionally, nothing is known about viruses infecting globe artichoke at subliminal levels or viral sequences embedded in the host genome, which are the newly recognized components of plant virome [8–10].

While the abundant in situ biodiversity of globe artichoke represents an invaluable genetic resource for breeders, industry, and consumers, at the same time, it is imperative to find a way to restore and maintain a sanitary status compatible with the conservation of such biodiversity. Plant nurseries offer a solution for the ex situ maintenance and provision of virus-free planting material to farmers and breeders because the current EU Directives 93/61/CEE and 93/62/CEE enforce nursery productions to be based on virus-free and true-to-type certified stocks.

Here we describe a sanitation protocol based on the combination of in vitro meristem-tip culture and thermotherapy for the production and the ex situ conservation of virus-free propagation material of five local varieties of globe artichoke, relevant to the Apulian RDP and in conformity to the EU Directives 93/61/CEE and 93/62/CEE.

Table 1. Viruses infecting globe artichoke in nature: taxonomic allocation, epidemiology, and disease symptoms.

Viruses with Isometric Particles	Genus Family	Vector/Mode of Transmission	Disease Symptoms
artichoke Aegean ringspot virus (AARSV)	*Nepovirus* (Subgroup A) *Comoviridae*	unknown	Yellow blotches and mild mottling. Often symptomless
artichoke Italian latent virus (AILV) [1]	*Nepovirus* (Subgroup B) *Comoviridae*	*Longidorus apulus* *L. fasciatus* Seed coat and expanded cotyledons	Mostly symptomless; patchy chlorotic stunting observed in some cvs
artichoke mottled crinkle virus (AMCV) [1]	*Tombusvirus* *Tombusviridae*	Contact Soil-borne	Severe deformations, mottling and crinkling of the leaves. Severe distortion of flower capitula. Plant death. New foliage emerging from underground buds develops poorly and often shows bright chrome-yellow discoloration
artichoke yellow ringspot virus (AYRSV)	*Nepovirus* (Subgroup C) *Comoviridae*	Pollen and seed transmission in tobacco and onion	Bright yellow blotches, ringspots and line patterns sometime followed by necrosis
artichoke vein banding virus (AVBV)	*Cheravirus* *Secoviridae*	unknown	Chlorotic discolorations along the leaf veins
broad bean wilt virus (BBWV) [1]	*Fabavirus* *Comoviridae*	*Capitophorus horni* and other aphid species	Yellow mottle, mosaic, or line patterns
cucumber mosaic virus (CMV) [1]	*Cucumovirus* *Bromoviridae*	several aphid species	Probably symptomless. Usually found in mixed infection with ArLV and/or TSWV in plants showing severe stunting
pelargonium zonate spot virus (PZSV) [1]	*Anulavirus* *Bromoviridae*	several thrips species	Moderate stunting and leaf chlorotic mottling. Chlorotic spots and line patterns after inoculation under experimental conditions
tobacco streak virus (TSV)	*Ilarvirus* (Subgroup 1) *Bromoviridae*	several thrips species	Stunting and leaf deformation
tomato black ring virus (TBRV)	*Nepovirus* (Subgroup B) *Comoviridae*	*Longidorus attenuatus*	Mild leaf mottling
Viruses with Rod-Shaped Particles	**Genus Family**	**Vector/Mode of Transmission**	**Disease Symptoms**
tobacco mosaic virus (TMV) [1]	*Tobamovirus*	contact	symptomless
tobacco rattle virus (TRV)	*Tobravirus*	*Thrichodorus christiei*	Leaf bright yellow discoloration

Table 1. *Cont.*

Viruses with Isometric Particles	Genus Family	Vector/Mode of Transmission	Disease Symptoms
Viruses with Filamentous Particles	**Genus Family**	**Vector/Mode of Transmission**	**Disease Symptoms**
artichoke curly dwarf virus (ACDV)	*Potexvirus* Flexiviridae	unknown	Stunting, leaf distortion and extended vein necrosis, delay in the development of flower capitula
Artichoke degeneration virus (ADV). Putative potyvirus	Non-classified	unknown	Leaf mottling curling and crinkling
Artichoke latent virus (ArLV)[1]	*Macluravirus* Potyviridae	*Myzus persicae, Brachycaudus cardui, Aphis fabae*	Symptomless
artichoke latent M virus (ArLMV)	*Carlavirus* Flexiviridae	unknown	Symptomless. Often in mixed infection with AILV, ArLV, AMCV
artichoke latent S virus (ArLSV)	*Carlavirus* Flexiviridae	aphids	Symptomless
bean yellow mosaic virus (BYMV)[1]	*Potyvirus* Potyviridae	several aphid species	Leaf yellowing, yellow flacking, line patterns, often in mixed infection with ArLV
Filamentous viruses and globe artichoke disease	*unknown*	unknown	Foliar mottling, leaf deformation and crinkling, stunting, and decreased yield. Progressive degeneration
potato virus X (PVX)[1]	*Potexvirus* Flexviridae	unknown	Mosaic, leaf deformation and narrowing, stunting
ranunculus latent virus (RaLV)			Mostly symptomless. Often in mixed infection with cynara 42 virus (Cy42)
turnip mosaic virus (TuMV)[1]	*Potyviridae Potyvirus*	several aphid species	Symptomless. Often in mixed infection with ArLV
tomato infectious chlorosis virus (TICV)[1]	*Crinivirus* Closteroviridae	*Trialeurodes vaporariorum*	Symptomless or mild interveinal yellowing
Unnamed putative carla virus	Non-classified	unknown	Symptomless
Viruses with Enveloped Particles	**Genus Family**	**Vector/Mode of Transmission**	**Disease Symptoms**
cynara virus (CraV). Putative cytorhabdovirus	unassigned *Rhabdoviridae*	unknown	Progressive degeneration. Often in mixed infection with AILV, ArLV, BBW, BYMV
tomato spotted wilt virus (TSWV)[1]	*Tospovirus Bunyaviridae*	several thrips species	severe stunting and deformation, generalized chlorosis and bronzing of the apical leaves, distortion of the head stalk, and necrosis of portions of the inner and outer scales

[1] Viruses that should be absent from nursery stocks, according to Directive 447-03/08/12 of the Apulian Regional Government.

Figure 1. Apulian artichoke ecotypes subjected to sanitation from virus infection by meristem-tip and thermotherapy (**a**) "Bianco di Taranto"; (**b**) "Francesina"; (**c**) "Locale di Mola"; (**d**) "Verde di Putignano"; (**e**) "Violetto di Putignano"; (**f**) Map of province sub-areas of Apulia region. (Photo (**a,b,d,e**) courtesy of Dr. Gabriella Sonnante, CNR IBBR collection, Bari, Italy; photo (**c**), courtesy of Prof. Pietro Santamaria, University of Bari Aldo Moro, Bari, Italy).

2. Materials and Methods

2.1. Plant Material and Assessment of the Sanitary Status

Offshoots of the Apulian early-flowering artichoke ecotypes "Francesina" (synonym "Violetto di Brindisi", "Violetto di San Ferdinando") "Verde di Putignano", "Violetto di Putignano" and "Locale di Mola" and of the late-flowering ecotype "Bianco di Taranto" (Figure 1) were collected from local farmers and maintained in a commercial nursery Plant as "standard initial material". Offshoots were tested for the presence of the following artichoke viruses considered economically relevant by the Directive 447-03/08/12 of the Apulian Regional Government (Table 1): artichoke Italian latent virus (AILV); artichoke latent virus (ArLV); artichoke mottled crinkle virus, (AMCV); bean yellow mosaic virus (BYMV); cucumber mosaic virus (CMV); pelargonium zonate spot virus (PZSV); tomato infectious chlorosis virus (TICV); tobacco mosaic virus (TMV); tomato spotted wilt virus (TSWV) and turnip mosaic virus (TuMV). Tests were carried out by dot-blot hybridization following the protocol of Minutillo et al. [11]. Chemiluminescent signals derived from the Digoxigenin-labelled DNA probes were acquired by exposing the membrane under a Chemi-Doc apparatus (Bio-Rad) and analyzed with Quantity One software (Bio-Rad).

2.2. Plant Sanitation Protocol

Young offshoots (10–15 cm) were collected from symptomless plants of each artichoke ecotype grown in situ by local farmers and transplanted into pots to be maintained ex situ in the nursery. Five shoots of each ecotype were used for the sanitation protocol (Figure 2). Each shoot was washed with tap water, disinfected with a 20% [v/v] of a commercial bleach solution (40 g·L^{-1} active chlorine) for 20 min, and then washed three times with sterile distilled water under a laminar flow hood.

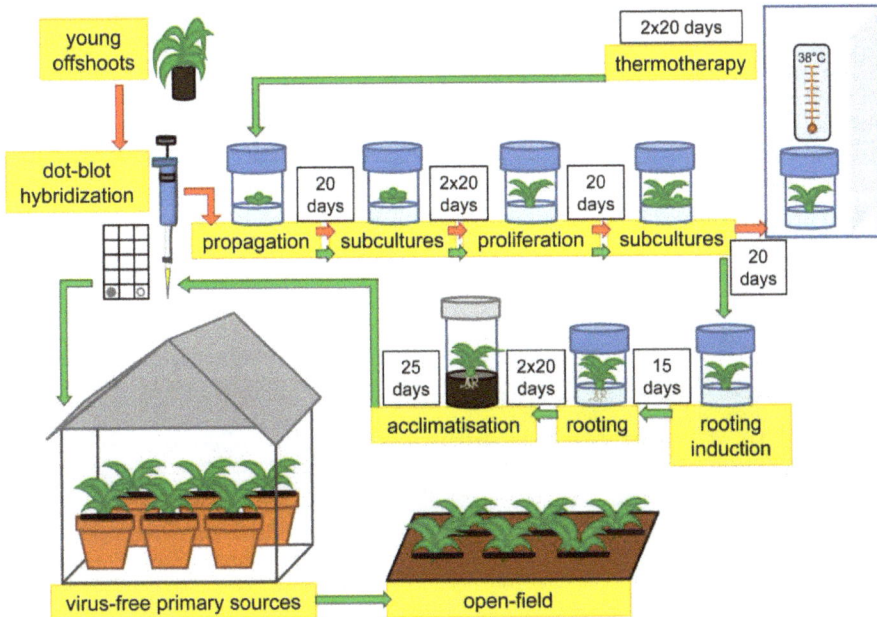

Figure 2. Schematic representation of sanitation workflow to produce virus-free artichoke plants using in vitro meristem-tip culture combined with thermotherapy. Red arrows indicate the workflow of sanitation protocol performed on young offshoots collected from farmers until thermotherapeutic treatment. Green arrows indicate the workflow of sanitation protocol performed after thermotherapy treatment.

Shoot meristem-tips (0.2–0.4 mm) were excised under a stereomicroscope and cultured into 60 mL polystyrene containers with 10 mL of propagation medium 1 (PM1), pH 5.8, consisting of a basal medium (BM) supplemented with 4.9 µM γ·γ dimethylallilminopurine (2ip), 2.8 µM gibberellic acid (GA3), 5.7 µM indoleacetic acid (IAA) and MS vitamins [12], to set up appropriate in vitro conditions [13]. The BM contained MS macronutrients, NN micronutrients [14], 68 µM Fe-EDTA, 2% [w/v] sucrose and 0.7% [w/v] agar.

Shoot meristem-tips were incubated at 24 ± 1 °C in a growth chamber with a 16-h light photoperiod at 4000 lx LED light intensity. Viable shoots were transferred to fresh PM1 at 20-day intervals. After three sub-cultures on PM1, young shoots (2–3 cm) were transferred into 200 mL glass culture vessels with 30 mL of proliferation medium 2 (PM2) containing BM, 0.46 µM Kinetin, 0.5 µM 1-naphthaleneacetic acid (NAA), Z4 vitamins [15], 5.6 mM ascorbic acid and 5.2 mM citric acid.

After two sub-cultures, lateral young shoots were transferred to PM1 and exposed at 38 ± 1 °C for two cycles of 20-days each, in a growth chamber with a 16-h light photoperiod at 5000 lx cool light intensity.

After thermotherapy, shoot meristem-tips were excised from explants that survived treatment and sub-cultured again on PM1 and PM2, as described above. Young shoots (3–4 cm) were transferred to 200 mL glass culture vessels containing 30 mL of rooting induction medium 1 (RM1), containing BM, 55.5 μM myo-inositol, 0.3 μM thiamine HCl, 0.4 mM adenine sulfate and 14.7 μM indole-3-butyric acid (IBA). After 15-day induction on RM1, rhizogenesis was obtained by transferring shoots on rooting medium 2 (RM2), containing BM supplemented with 55.5 μM myo-inositol, 0.3 μM thiamine HCl and 57 μM IAA.

The rooted micro-rosettes were transplanted to pre-compressed peat disks (Jiffy-7® pellet) soaked with a solution of 7.4 μM IBA and grown at 23 ± 1 °C in an acclimatization room with 16-h light photoperiod at 3000 lx LED light intensity. Over the 20–25 days of the acclimatization phase, relative humidity (RH) was gradually reduced from 85–90% to 55–60%. Acclimatized plants were transferred to a greenhouse and grown at 18–20 °C, 55–60% RH and 16-h cool light photoperiod.

Plants grown in the greenhouse were tested for virus presence by dot-blot hybridization, as described above and by reverse-transcription polymerase chain reaction (RT-PCR) using primer pairs and protocols described by Minutillo et al. [11]. Virus-free clones were transferred to the nursery to set up a collection of virus-free "primary sources" (Figure 3). After six months maintenance and multiplication in the nursery Plant, primary sources were checked again for virus presence by dot-blot hybridization.

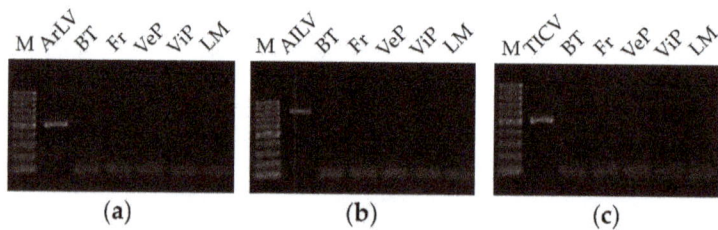

Figure 3. RT-PCR performed with primer pairs specific for (**a**) ArLV; (**b**) AILV and (**c**) TICV on nucleic acid extracted from primary sources of the five ecotypes of globe artichoke (BT, Bianco di Taranto; Fr, Francesina; VeP, Verde di Putignano; ViP, Violetto di Putignano; LM, Locale di Mola).

3. Results

Results from dot blot hybridization revealed that only the offshoots collected from the ecotype "Francesina" did not harbor infection of any of the viruses tested, while the other ecotypes considered in this study had single or mixed infections by ArLV, AILV and TICV (Table 2). Nonetheless, all the ecotypes were included in the sanitation protocol. Only 36% of the young offshoots derived from meristem tips were stabilized in vitro. The rest of the stock became contaminated by fungi, which had not been removed by the sterilization procedure.

After two cycles of sub-culture in PM2 the number of explants increased with a multiplication rate (MR) ranging from 2.8 to 7 with a mean value of 3.74 ± 1.8, allowing the obtaining of a sufficient number of shoots to enter the next phase of in vitro thermotherapy. Some of the explants died during the treatment while an average of 75% of micro-rosettes survived and entered the second cycle of meristem-tip culture. During the second cycle the number of explants increased with a multiplication rate (MR) that was very similar to the MR recorded in the first phase of the sanitation protocol and mostly similar among the ecotypes tested. On the contrary, rooting capacity (RC) was extremely variable (Table 2). There was an inverse relationship between MR and RC as to a higher MR corresponded a reduced RC. This problem was partially resolved by the addition of 14.7 μM IBA in RM1 followed by 57 μM IAA in RM2 that gave acceptable root and shoot regeneration for all the ecotypes. In our study, the ecotype "Verde di Putignano" showed the highest RC (89.7%) and a MR of only 2.7. Conversely, the ecotype "Locale di Mola" had the highest MR (7) and a RC of 15.4%. The ecotypes "Bianco di Taranto" and "Violetto di Putignano" showed intermediate results while the ecotype and Francesina had the, lowest RC (Table 2).

Finally, our results showed that plant survival during acclimatization phase was clearly affected by the robustness of the radical apparatus. Eighty-five percent of the rooted plants acclimatized successfully, whereas only 10% of the poorly rooted or unrooted plants were able to survive in greenhouse conditions. After the completion of sanitation protocol and acclimatization (Figure 2), we obtained a total of 25 primary sources (Table 2) that resulted virus-free when tested by dot blot hybridization and RT-PCR (Figure 3). This condition was maintained after a six-month culture and multiplication in the nursery when all the plants were checked again (Figure 4). After multiplication the number of primary sourced raised to 361 plants of Bianco di Taranto, 273 plants of Violetto di Putignano, 78 plants of Verde di Putignano, 69 plants of Francesina and 98 plants of Locale di Mola.

Table 2. Viruses detected in the off-shoots of globe artichoke ecotypes collected from farmers and their multiplication rates (MR) and rooting capacity (RC) obtained during the sanitation protocol.

Ecotype	Standard Material [1]	Virus Detected [2]	Stabilized Shoots [3]	MR [4]	RC [5]	Primary Sources [6]
Bianco di Taranto	5	ArLV	3	2.8	68.9%	6
Francesina	5	-	1	3	12.5%	1
Verde di Putignano	5	TICV	2	2.7	89.7%	9
Violetto di Putignano	5	TICV	1	3.2	63.6%	7
Locale di Mola	5	ArLV, AILV	2	7	15.4%	2

[1] Number of initial offshoots of each ecotype that entered the sanitation protocol; [2] Viruses detected by Dot-blot hybridization in each offshoot of standard material; [3] Number of shoots obtained after three sub-culture cycles in PM1; [4] MR after two sub-culture cycles in PM2; [5] RC after two sub-culture cycles in RM2; [6] Total number of primary sources obtained for each ecotype at the end of the sanitation protocol.

(a)   (b)

Figure 4. (a) Virus-free primary sources of the five ecotypes of globe artichoke maintained in the nursery "F.lli Corrado S.r.L." Greenhouse lateral openings and doors are protected with aphid-proof screens to preserve the phytosanitary condition obtained with the sanitation protocol; (b) Clonal in vivo multiplication of primary sources.

4. Discussion

The detection of mixed infections in symptomless artichoke plants confirmed that the selection of offshoots from these plants for vegetative multiplication does not guarantee the absence of virus infections. The occurrence of TICV infection in the two varieties collected from the Putignano area in the province of Bari is relevant from the eco-epidemiological point of view (Figure 1). TICV is an emerging pathogen [16,17] transmitted semi-persistently by the whitefly *Trialeurodes vaporariorum*, but not by *Bemisia tabaci*. Previous investigations in Italy detected TICV in tomato and globe artichoke in Liguria, Sardinia, Latium, Campania, and Sicily. Artichoke plants infected by TICV did not show clear symptoms or, at most, they showed interveinal yellowing of the leaves but in all instances, plants were infested with high populations of *T. vaporariorum* [18,19] that ensured transmission of the virus to other susceptible crops. In Apulia, the virus was recorded for the first time in mixed infection with TSWV in a tomato greenhouse in the Lecce province [20] and a second time in 2015 in another tomato greenhouse in the Bari province [21]. Interestingly, the two ecotypes "Violetto di Putignano" and "Verde di Putignano" infected by TICV were grown in the same area of the Bari province and infections may have been the result of intense fluctuations of *T. vaporariorum*, which are very common. Such symptomless artichoke plants escape roguing and may act as perennial TICV reservoirs from where populations of *T. vaporariorum* can acquire the virus and cause outbreaks on tomato.

The sanitation protocol was applied to all the offshoots collected. During the experimental procedure, we identified three critical steps. The first critical phase was the in vitro stabilization of meristem-tips cut from the young offshoots because only 36% of them were stabilized in vitro and used for the culture. The high rate of contamination observed were probably due to vascular fungi localized internally. This initial loss of explants was offset by a high multiplication rate that was considered satisfactory and in the range of 3.7–4.8 new shoots per explant obtained in other studies ([22] and references quoted therein). Based on previous experience [23], we did not proceed with further subcultures to reduce the risk for the induction of "pastel variants" and loss of earliness [24] for the early-flowering ecotypes.

Generally, in vitro thermotherapy is the second critical step, but compared to previous experience [23] a higher rate of micro-rosettes survived to the treatment and produced new shoots suitable for the second cycle of meristem-tip culture. This was probably due to the more stringent control of light, temperature, and RH parameters in comparison to the equipment and facilities used in the previous study. Thermotherapy proved to be an essential step to eradicate plant viruses owing to their uneven distribution in plant tissues and to the notion that some viruses may also colonize meristem-tips [25–28]. Virus localization in meristem-tips seems particularly pertinent to nepoviruses [29] and to AILV among the viruses infecting globe artichoke [30]. A markedly enhanced virus eradication by the combined action of meristem-tip culture and thermotherapy was also reported in other instances [28,31,32].

The third critical step was the induction of root formation. It was reported that MR and RC require different auxin/cytokinin ratios. High ratios of auxin/cytokinin promote root regeneration, while high ratios of cytokinin/auxin stimulate shoot regeneration [33]. In our tests, a substantially balanced of auxin/cytokinin in PM1 and PM2 produced a satisfactory rate of root and shoot regeneration for all the ecotypes. López-Pérez and Martínez [22] observed a higher root induction rate of globe artichoke when a high concentration of IBA in the culture medium associated with 5 days of darkness was used to induce the rhizogenesis. In our conditions a high auxin concentration in RM steps favored rhizogenesis whereas 5 days of darkness did not induce any ameliorative effect. Benoit and Ducreux [34] and Morzadec and Hourmant [35] reported the process of root induction to be also highly dependent from the ecotype and may range from 1 to 92%.

The application of the sanitation protocol allowed the obtaining of a sufficient number of virus-free primary resources to be used for nursery activity. These were increased 50-fold after one year of rapid multiplication in the nursery plant. Only the ecotype Verde di Putignano increased 9-fold. The results confirmed those of a previous study, i.e., that the combined action of meristem-tip culture and thermotherapy is required to eliminate also the infection of AILV, which localizes in the meristem-tip through which it is probably transmitted to seeds [30,36].

5. Conclusions

Regional ecotypes of globe artichoke represent a valuable genetic resource that need adequate handling and protection. This study confirmed that some of such ecotypes harbor virus infections that may have contributed to their progressive degeneration and abandonment. The sanitation protocol had a more balanced auxin/citokinin ratio and better standardized parameters for in vitro growing and thermotherapy compared to that used for the reflowering type Brindisino [23]. We did not evaluate the effect of sanitation on the earliness or other agronomic traits of the ecotypes sanitized with a dedicated experimental field design, but a very preliminary small-scale field test shows promising results (Figure 5).

(a) (b)

Figure 5. Production of sanitized clones of ecotypes "Locale di Mola" (**a**) and "Violetto di Putignano" (**b**) in a small-scale experimental field for a preliminary evaluation of agronomic traits after sanitation.

The keystone to conservation of such ecotypes relies on the maintenance of the phytosanitary status obtained with sanitation. In situ maintenance would be ideal to preserve the ecological niche but it is a matter of fact that if exposed to field condition the sanitized material will undergo reinfection quickly. The reflowering type Brindisino showed an estimated ArLV reinfection rate between 1% and 2% after 2 years from planting close to a standard artichoke field with ArLV incidence higher than 80% [7].

Therefore, sanitized stocks must be maintained ex situ in nursery structures adequately protected to prevent reinfection and must be subjected periodically to phytosanitary controls.

Available diagnostics possess sufficient sensitivity and specificity to detect even subliminal infections provided that the etiologic agent is known. We already briefly addressed the issue of the complexity of plant viromes that is emerging from the application of NGS technology. Besides the biological significance, the discovery of a previously undescribed virus species in asymptomatic plants will impact quarantine regulations and issues for the virus-free certified stocks [37–40]. This issue must be addressed to fulfill expectations from the conservation and use of agricultural biodiversity and genetic resources [41]. It is expected that the use of sanitized ecotypes supplied by nursery-certified stocks and cultivated in open field for no more than two years should decrease inoculum potential and will ensure genetic resources for conservation, breeding and production.

Acknowledgments: This work was supported by the Regional Apulian project "Biodiversity of Apulian vegetable species" CE n. 1698/2005—Programma di Sviluppo Rurale per la Puglia 2007–2013. Misura 214–Azione 4 Sub azione a): "Progetti integrati per la biodiversità" grant CUP H98B13000000005, Italy. We acknowledge Donato Gallitelli, University of Bari aldo Moro, for critically reading the manuscript and Giovanni Corrado, Nursery Plant Vivaio F.lli Corrado S r L , Torre S. Susanna (BR), Italy for collecting and maintaining ex situ the artichoke ecotypes used in this study.

Author Contributions: R.S., G.B. and T.M. conceived and designed the experiments; R.S., A.Co. and A.Ca. performed the experiments; R.S., A.D.F. and G.B. analyzed the data; R.S. and T.M. wrote the paper.

Conflicts of Interest: The authors declare no conflict of interest. The founding sponsors had no role in the design of the study; in the collection, analyses, or interpretation of data; in the writing of the manuscript, and in the decision to publish the results.

References

1. Romani, A.; Pinelli, P.; Cantini, C.; Cimato, A.; Heimler, D. Characterization of Violetto di Toscana, a typical Italian variety of artichoke (*Cynara scolymus* L.). *Food Chem.* **2006**, *95*, 221–225. [CrossRef]
2. Mauro, R.P.; Portis, E.; Acquadro, A.; Lombardo, S.; Mauromicale, G.; Lanteri, S. Genetic diversity of globe artichoke landraces from Sicilian small-holdings: Implications for evolution and domestication of the species. *Conserv. Gene* **2009**, *10*, 431–440. [CrossRef]
3. Mauro, R.P.; Portis, E.; Lanteri, S.; Mauromicale, G. Genotypic and bio-agronomical characterization of an early Sicilian landrace of globe artichoke. *Euphytica* **2012**, *186*, 357–366. [CrossRef]
4. Ciancolini, A.; Rey, N.A.; Pagnotta, M.A.; Crinò, P. Characterization of Italian spring globe artichoke germplasm: Morphological and molecular profiles. *Euphytica* **2012**, *186*, 433–443. [CrossRef]
5. Tavazza, R.; Lucioli, A.; Benelli, C.; Giorgi, D.; D'Aloisio, E.; Papacchioli, V. Cryopreservation in artichoke: Towards a phytosanitary qualified germplasm collection. *Ann. Appl. Biol.* **2013**, *163*, 231–241. [CrossRef]
6. Elia, A.; Santamaria, P. Biodiversity in vegetable crops, a heritage to save: The case of Puglia region. *Ital. J. Agron.* **2013**, *8*, 21–34. [CrossRef]
7. Gallitelli, D.; Mascia, T.; Martelli, G.P. Viruses in Artichoke. *Adv. Virus Res.* **2012**, *84*, 289–324. [PubMed]
8. Virgin, H.W. The virome in mammalian physiology and disease. *Cell* **2014**, *157*, 142–150. [CrossRef] [PubMed]
9. Tollenaere, C.; Susi, H.; Laine, A.L. Evolutionary and epidemiological implications of multiple infection in plants. *Trends Plant Sci.* **2015**, *21*, 80–90. [CrossRef] [PubMed]
10. Mascia, T.; Gallitelli, D. Synergies and antagonisms in virus interactions. *Plant Sci.* **2016**, *252*, 176–192. [CrossRef] [PubMed]
11. Minutillo, S.A.; Mascia, T.; Gallitelli, D. A DNA probe mix for the multiplex detection of ten artichoke viruses. *Eur. J. Plant Pathol.* **2012**, *134*, 459–465. [CrossRef]
12. Murashige, T.; Skoog, F. A revised medium for rapid growth and bioassays with tobacco tissue cultures. *Physiol. Plant.* **1962**, *15*, 473–497. [CrossRef]
13. Fortunato, I.M.; Ruta, C.; Castrignano, A.; Saccardo, A. The effect of mycorrhizal symbiosis on the development of micropropagated artichokes. *Sci. Hortic.* **2005**, *106*, 472–483. [CrossRef]
14. Nitsch, J.P.; Nitsch, C. Haploid plants for pollen grains. *Science* **1969**, *63*, 85–87. [CrossRef] [PubMed]
15. Lloyd, G.; McCown, B. Commercially-feasible micropropagation of mountain laurel, Kalmia latifolia, by use of shoot-tip culture. *Int. Plant Prop. Soc. Proc.* **1980**, *30*, 421–427.
16. Hanssen, I.M.; Lapidot, M.; Thomma, P.H.J.B. Emerging viral diseases of tomato crops. *Mol. Plant Microbe Interact.* **2010**, *23*, 539–548. [CrossRef] [PubMed]
17. Hanssen, I.M.; Lapidot, M. Major tomato viruses in the Mediterranean basin. *Adv. Virus Res.* **2012**, *84*, 31–66. [PubMed]
18. Vaira, A.M.; Accotto, G.P.; Vecchiati, M.; Bragaloni, M. Tomato infectious chlorosis virus causes leaf yellowing and reddening of tomato in Italy. *Phytoparasitica* **2002**, *30*, 290–294. [CrossRef]
19. Davino, S.; Tomassoli, L.; Tiberini, A.; Mondello, V.; Davino, M. Outbreak of tomato infectious chlorosis virus in a relevant artichoke producing area of Sicily. *J. Plant Pathol.* **2009**, *91*, S4.57–S4.58.
20. Spanò, R.; Mascia, T.; Minutillo, S.A.; Gallitelli, D. First report of Tomato infectious chlorosis virus from tomato in Apulia, southern Italy. *J. Plant Pathol.* **2011**, *93*. [CrossRef]
21. Spanò, R.; Trisciuzzi, N.; Gallitelli, D. Gravi infezioni di Tomato infectious chlorosis virus su pomodoro in Puglia. *Prot. Colt.* **2015**, *4*, 43–45.
22. López-Pérez, A.J.; Martínez, J.A. In vitro root induction improvement by culture in darkness for different globe artichoke cultivars. *In Vitro Cell. Dev. Biol. Plant* **2015**, *51*, 160–165. [CrossRef]
23. Acquadro, A.; Papanice, M.A.; Lanteri, S.; Bottalico, G.; Portis, E.; Campanale, A.; Finetti-Sialer, M.M.; Mascia, T.; Sumerano, P.; Gallitelli, D. Production and fingerprinting of virus-free clones in a reflowering globe artichoke. *Plant Cell Tissue Organ Cult.* **2010**, *100*, 329–337. [CrossRef]
24. Pecaut, P.; Martin, F. Variation occuring after natural and in vitro multiplication of early Mediterranean cultivars of globe artichoke (*Cynara scolymus* L.). *Agronomie* **1993**, *13*, 909–919. [CrossRef]
25. Walkey, D.G.A.; Webb, M.J. Virus in plant apical meristem. *J. Gen. Virol.* **1968**, *3*, 311–313. [CrossRef]
26. Appiano, A.; Pennazio, S. Electron microscopy of potato meristem tips infected with potato virus X. *J. Gen. Virol.* **1972**, *14*, 273–276. [CrossRef] [PubMed]

27. Mori, K.; Hosokawa, D. Localization of viruses in apical meristem and production of virus-free plants by means of meristem and tissue culture. *Acta Hortic.* **1977**, *78*, 389–396. [CrossRef]

28. Wang, Q.; Cuellar, W.J.; Rajamäki, M.L.; Hirata, Y.; Valkonen, J.P.T. Combined thermotherapy and cryotherapy for efficient virus eradication: Relation of virus distribution, subcellular changes, cell survival and viral RNA degradation in shoot tips. *Mol. Plant Pathol.* **2008**, *9*, 237–250. [CrossRef] [PubMed]

29. Goshal, B.; Sanfaçon, H. Symptom recovery in virus-infected plants: Revisiting the role of RNA silencing mechanisms. *Virology* **2015**, *479–480*, 167–179. [CrossRef] [PubMed]

30. Santovito, E.; Mascia, T.; Siddiqui, S.A.; Minutillo, S.A.; Valkonen, J.P.T.; Gallitelli, D. Infection cycle of artichoke italian latent virus in tobacco plants: Meristem invasion and recovery from disease symptoms. *PLoS ONE* **2014**, *9*. [CrossRef] [PubMed]

31. Walkey, D.G.A.; Cooper, V.C. Effect of temperature on virus eradication and growth of infected tissue cultures. *Ann. Appl. Biol.* **1975**, *80*, 185–190. [CrossRef] [PubMed]

32. Cooper, V.C.; Walkey, D.G.A. Thermal inactivation of cherry leaf roll virus in tissue cultures of Nicotiana rustica raised from seeds and meristem tips. *Ann. Appl. Biol.* **1978**, *88*, 273–278. [CrossRef]

33. Tang, L.P.; Li, X.M.; Dong, Y.X.; Zhang, X.; Sand Su, Y.H. Microfilament depolymerization is a pre-requisite for stem cell formation during in vitro shoot regeneration in Arabidopsis. *Front. Plant. Sci.* **2017**, *8*, 158. [CrossRef] [PubMed]

34. Benoit, H.; Ducreux, G. Etude de quelques aspects de la multiplication végétative in vitro de l'artichaut (*Cynara scolymus* L.). *Agronomie* **1981**, *1*, 225–230. [CrossRef]

35. Morzadec, J.M.; Hourmant, A. In vitro rooting improvement of globe artichoke (cv Camus de Bretagne) by GA3. *Sci. Hortic.* **1997**, *72*, 59–62. [CrossRef]

36. Bottalico, G.; Padula, M.; Campanale, A.; Finetti Sialer, M.M.; Saccomanno, F.; Gallitelli, D. Seed transmission of Artichoke Italian latent virus and Artichoke latent virus in globe artichoke. *J. Plant Pathol.* **2002**, *84*, 167–168.

37. Massart, S.; Olmos, A.; Jijakli, H.; Candresse, T. Current impact and future directions of high throughput sequencing in plant virus diagnostics. *Virus Res.* **2014**, *188*, 90–96. [CrossRef] [PubMed]

38. Massart, S.; Candresse, T.; Gil, J.; Lacomme, C.; Predajna, L.; Ravnikar, M.; Reynard, J.S.; Rumbou, A.; Saldarelli, P.; Škorić Vainio, E.J.; et al. A framework for the evaluation of biosecurity, commercial, regulatory, and scientific impacts of plant viruses and viroids identified by NGS technologies. *Front. Microbiol.* **2017**, *8*, 45. [CrossRef] [PubMed]

39. Wu, Q.; Ding, S.W.; Zhang, Y.; Zhu, S. Identification of viruses and viroids by next-generation sequencing and homology-dependentand homology-independent algorithms. *Annu. Rev. Phytopathol.* **2015**, *53*, 425–444. [CrossRef] [PubMed]

40. Martin, R.R.; Constable, F.; Tzanetakis, I.E. Quarantine regulations and the impact of modern detection methods. *Annu. Rev. Phytopathol.* **2016**, *54*, 189–205. [CrossRef] [PubMed]

41. Esquinas-Alcázar, J. Protecting crop genetic diversity for food security: Political, ethical and technical challenges. *Nat. Rev. Genet.* **2005**, *6*, 946–953. [CrossRef] [PubMed]

agriculture

MDPI

Article

The Deterioration of Morocco's Vegetable Crop Genetic Diversity: An Analysis of the Souss-Massa Region

Stuart Alan Walters [1,*], Rachid Bouharroud [2], Abdelaziz Mimouni [2] and Ahmed Wifaya [2]

[1]   Department Plant, Soil, and Agricultural Systems, Southern Illinois University, Carbondale, IL 62901 USA
[2]   Integrated Crop Production Unit, National Institute of Agronomic Research, Regional Center of Agadir, Agadir 80350, Morocco; bouharroud@yahoo.fr (R.B.); mimouniabdelaziz@yahoo.fr (A.M.); wifaya_ahmed@yahoo.fr (A.W.)
*   Correspondence: awalters@siu.edu; Tel.: +1-618-4532496

Received: 7 March 2018; Accepted: 28 March 2018; Published: 30 March 2018

Abstract: Crop domestication and breeding efforts during the last half-century in developed countries has significantly reduced the genetic diversity in all major vegetable crops grown throughout the world. This includes developing countries such as Morocco, in which more than 90% of all farms are less than 10 ha in size, which are generally maintained by subsistence farmers who try to maximize crop and animal productivity on a limited land area. Near Agadir, in the remote Anti-Atlas mountain areas of the Souss-Massa region, many small landowner vegetable growers are known to still utilize crop populations (landraces). Thus, an assessment of the current status of vegetable landraces was made in this mountainous region of Southwestern Morocco during 2014. This assessment indicated that a significant loss of vegetable crop landraces has occurred in the last 30 years in this region of Morocco. Although many vegetable crops are still maintained as landrace populations by small subsistence farmers in remote areas in the Souss-Massa region, only 31% of these farmers cultivated landraces and saved seed in the villages assessed, with the average farmer age cultivating landraces being 52 years old. Moreover, the approximated loss of vegetable crop landraces over the last 30 years was an astounding 80 to 90%. Vegetable crops notably lost during this time period included carrot (*Daucus carota*), fava beans (*Vicia faba*), melon (*Cucumis melo*), pea (*Pisum sativum*), watermelon (*Citrullus lanatus*), and tomato (*Solanum lycopersicon*). The most significant loss was tomato as no landraces of this crop were found in this region. The vegetable crop landraces that are still widely grown included carrot, melon, onion (*Allium cepa*), turnip (*Brassica rapa* var. *rapa*), and watermelon, while limited amounts of eggplant (*Solanum melongea*), fava bean, pea, pepper (*Capsicum annuum*), and pumpkin (*Cucurbita moshata* and *C. maxima*) were found. This recent genetic deterioration will have a profound influence on future Moroccan agricultural productivity, as the genetic diversity within these landraces may be the only resource available to allow these smaller subsistence farmers to cope with changing environmental conditions for the optimization of crop production in their harsh climate.

Keywords: crop population; genetic resources; genetic variability; germplasm; landrace

1. Introduction

The domestication of improved vegetable genotypes through breeding efforts during the last half-century has significantly reduced the genetic diversity in all major vegetable crops grown throughout the world. Growers and consumers alike have become accustomed to the high quality and yields of hybrid vegetable cultivars, which has definitely resulted in a significant reduction in cultivation of landraces (or local varieties). Landraces are dynamic populations of a genetically-diverse, locally-adapted cultivated plant species that have historical origins, distinct identity, no formal crop improvement, and are oftentimes associated with traditional farming systems [1]. Although landraces generally have high genetic diversity and heterogeneity, they oftentimes will provide inconsistent phenotypes with overall lower quality and yields compared to newer hybrids or pure-line inbreds.

Vegetable crops grown in sustainable, low-input, traditional farming systems have traditionally been landraces. However, the cultivation of vegetable landraces has decreased since the widespread use of improved hybrid cultivars [2]. Since the first modern hybrid varieties were released more than 50 years ago, the areas planted to these improved varieties continue to expand [3]. Vegetable crop landrace cultivation has also diminished even in remote areas in developing countries, especially within the last few decades [4]. Landraces throughout the world are rapidly being lost, due to increasingly more limited use by growers in many areas of the world [2,3]; and, lower amounts of field cultivation, multiplication, and conservation by growers directly relates to the loss of this valuable genetic resource over time. Fewer and fewer subsistence farmers are cultivating landraces and maintaining seed stock from year to year. The loss of landrace populations is devastating to the vegetable crop species gene pool, as genes that have evolved over millennia are being lost. The main contributions of landraces to the development of new crop varieties have been traits for more efficient nutrient uptake and utilization, as well as genes for adaptation to stressful environments, such as water stress, salinity, and high temperatures [5]. Thus, landraces are an important diverse genetic resource that allows crop species to adapt to local effects of climate and pests, and to eventually produce improved yields unique to a specific environment, especially those that impose some type of stressful conditions on the growing crops. Crop genetic resources are crucial for the future survival of humanity, and future food security depends upon their conservation [6,7].

Morocco is a rapidly developing country and has great agricultural export markets in place due to its proximity to Europe. However, more than 90% of all farms are less than 10 ha, and these farms produce most of the vegetable crops that feed the populace [4]. There are some larger-sized farms that produce vegetables, but these tend to be financed from developed countries in Europe, especially France and Spain. Similar to other developing countries, income is very limited in Morocco (especially in rural areas) and most farmers must try to maximize productivity on a limited amount of land while trying to use minimal inputs. The Souss-Massa region of Morocco is an important agricultural area within this country, as it produces significant amounts of vegetables for both domestic and export sales. Agadir is the capital of this region, which is located in the southwestern portion of the country along the Atlantic coast. The villages immediately surrounding the city to the south are well known for their large-scale vegetable production operations, with products (especially greenhouse tomatoes) extensively grown for export to European markets. However, in remote areas of the Souss-Massa (especially in mountainous areas), most vegetable growers are subsistence farmers who minimize inputs to grow vegetable crops and typically sell their produce at local weekly markets (or souks). In past years, these small subsistence growers in rural areas of the Souss-Massa primarily relied upon seed from vegetable landraces for their crop production needs. The use of vegetable landraces resulted primarily due to the remoteness of these farmers, as well as the lack of contact to the outside world, which prevented access to other seed sources. However, the world has drastically changed in the last few decades and, today, there is readily available access to anything that anyone with money would desire even in remote areas of the world. Thus, the relatively small returns that landraces provide compared to the apparently overwhelming advantages that modern varieties seem to offer farmers are oftentimes not enough to maintain their utilization in a particular farming system [8]. Therefore, the objective of this study was to conduct an assessment of vegetable landraces in mostly remote areas within the Souss-Massa region of Morocco.

2. Materials and Methods

2.1. Study Area

This landrace assessment was conducted in the Souss-Massa region of Morocco. This area includes Agadir, a city of approximately 750,000 people, which has significant Berber influence, with an arid to semi-arid climate. Temperatures range from highs of ~40 to 45 °C in July and August to lows of ~2 to 4 °C in January. There is minimal rainfall with only 150 to 250 mm received annually, with little occurring from May to September, which is typically a very dry time of the year. More than 90% of the farmers in this region are small landholders, especially those in remote areas of the Anti-Atlas Mountains, and generally have only between 5 and 10 ha. This environment is oftentimes dry and hot which provides significant stress for plant growth and survival. The landscape of a typical, small subsistence farm in the Anti-Atlas mountains of Southwestern Morocco (near the village of Tizi N' Test) is provided in Figure 1.

Figure 1. Landscape of a typical, small subsistence farm in Anti-Atlas mountain region of Southwestern Morocco. Notice the onions growing in the basin irrigation beds underneath the olive trees.

2.2. Vegetable Landrace Assessment and Data Collection

During April 2014 a vegetable landrace assessment was conducted in the Souss-Massa region of Morocco. Six localities were visited to assess landraces in this region, and all were between latitudes 29.1620145° and 30.531266° N, and longitudes −9.636919° and −7.923879° W. The villages visited during the assessment were: Anou eljdid (30.162045, −9.252229); Biougra (30.220658, −9.372859); Massa (30.004087, −9.636919); Targa N' Touchka (29.885070, −9.199597); Taliouine (30.531266, −7.923879); and, Tizi N' Test (30.868223, −8.379130). The approximate location of each village in Southwestern Morocco is provided in Figure 2. Farmers were identified in each village and asked specific questions regarding landrace cultivation (Table 1). They provided answers to: total number of vegetable growers in each village, percentage of growers cultivating vegetable landraces and saving seed each year, average age of famers cultivating landraces, estimated percentage of vegetable crop landraces lost during the last 30 years, current vegetable landraces cultivated, and vegetable crop species landraces lost during the last 30 years. Seed were collected from the landraces that the farmers would provide to us and placed at the National Institute for Agronomic Research (INRA) in Settat, Morocco.

2.3. Data Analysis/Presentation

The data collected for the vegetable landrace assessment conducted in the Souss-Massa region of Morocco were compiled and placed into tabular form. The specific questions asked of famers in each village related to vegetable crop landraces were collated with overall means also determined for this data. Additional information collected regarding the vegetable crop landraces currently grown, as well as those lost over the last 30 years is also presented.

Figure 2. Approximate locations of the six villages in the Souss-Massa region of Morocco that were assessed for vegetable landrace use. The latitude (°N) and longitude (°W) of each village: Anou eljdid (30.162045, −9.252229); Biougra (30.220658, −9.372859); Massa (30.004087, −9.636919); Taliouine (30.531266, 7.923879); Targa N' Touchka (29.885070, −9.199597); and Tizi N' Test (30.868223, −8.379130).

Table 1. Questions provided to small subsistence farmers cultivating vegetable crop landraces in six villages within the Souss-Massa region of Morocco in 2014.

Questions:
What are the total number of vegetable growers in your village?
What is the percentage of growers cultivating vegetable landraces and saving seed each year in your village?
What is the average age of famers cultivating vegetable landraces in your village?
What is the estimated percentage of vegetable crop landraces that you think has been lost during the last 30 years?
What are the vegetable landraces currently cultivated in your village?
Which vegetable crop species landraces have been lost during the last 30 years?
The latitude (°N) and longitude (°W) of each village assessed: Anou eljdid (30.162045, −9.252229); Biougra (30.220658, −9.372859); Massa (30.004087, −9.636919); Taliouine (30.531266, −7.923879); Targa N' Touchka (29.885070, −9.199597); and Tizi N' Test (30.868223, −8.379130). Estimates for data presented in table based on conversations with small vegetable growers in each village.

3. Results

3.1. Farmer Demographics Related to Vegetable Landraces

The assessment indicated that there are relatively few small, subsistence farmers in the Souss-Massa region of Morocco that maintain vegetable landrace populations (Table 2). In the villages assessed, only about 31% of the small growers cultivated landraces and saved seed, although the range was from a low of 20% in Tizi N' Test to a high of 45% in Targa N' Touchka. This trend is disturbing, as growers are placing more reliance on other types of seed compared to landraces, which can have disastrous results for future vegetable crop production due to the increased lack of diversity that is being perpetuated in these production systems.

Table 2. Assessment of small, subsistence farmers cultivating vegetable crop landraces in six villages within the Souss-Massa region of Morocco in 2014, and estimated percentage of landraces lost in those villages over the last 30 years.

Moroccan Souss-Massa Region Village	Total No. of Growers in Village	Percent of Growers Cultivating Landraces and Saving Seed	Farmer Age Cultivating Landraces	Percent Landraces Lost in Last 30 Years
Anou eljdid	8	37	50	80
Biougra	8	25	52	90
Massa	15	33	55	90
Taliouine	13	23	45	80
Targa N' Touchka	20	45	55	80
Tizi N' Test	10	20	55	90

The latitude (°N) and longitude (°W) of each village assessed: Anou eljdid (30.162045, −9.252229); Biougra (30.220658, −9.372859); Massa (30.004087, −9.636919); Taliouine (30.531266, −7.923879); Targa N' Touchka (29.885070, −9.199597); and Tizi N' Test (30.868223, −8.379130). Estimates for data presented in table based on conversations with small vegetable growers in each village.

The average age of farmers cultivating landraces was 52 years old, and ranged between 45 in Taliouine, 50 in Anou eljdid, 52 in Biougra, and 55 in Massa, Targa N' Touchka, and Tizi N' Test (Table 2). The small number of growers cultivating vegetable landraces along with aging farmer populations have most likely directly resulted in the loss of landraces over time. This assessment indicated that there are few young farmers involved in the vegetable cultivation and the seed-saving process in the Souss-Massa region of Morocco, which is similar to many other areas of the world in which the younger generation has little to no interest in farming as a profession.

The estimated loss of vegetable crop landraces over the last 30 years by farmers in the villages assessed was an astounding 85% (Table 2). Three villages, Biougra, Massa, and Tizi N' Test, had experienced the loss of 90% of vegetable landraces over the last 30 years, while the other villages assessed was estimated at 80%. This estimated loss of vegetable crop diversity in this region of Morocco is much more than expected, and this trend is expected to continue in the near future.

3.2. Vegetable Crops as Landraces

Numerous vegetable crop landraces have been lost during the last 30 years in the Souss-Massa region of Morocco (Table 3). The complete loss of tomato landraces during the last 30 years was entirely unexpected, as five of the six villages assessed (or 83%) had recently grown landraces of tomato that are now totally lost. Other significant vegetable crop landrace losses in Souss-Massa villages during this time period were melon (67%), watermelon and fava bean (50%), and carrot and pea (17%). The significant losses of vegetable crop landraces in these villages during the last 30 years were expected, but not at the high levels that were determined from this assessment.

Although many vegetable crop landraces have declined over the last 30 years, many are still widely grown in this region, including carrot, melon, onion, pumpkin, turnip, and watermelon (Table 3). There were limited amounts of other vegetable landraces cultivated in this region, and included eggplant, fava bean, pea, and pepper; these were all found in Tizi N' Test, as none of the other locations evaluated grew landraces of these crops. Melon and onion landraces were the most prevalent throughout this region of Morocco, with 67% of the villages assessed currently growing landraces of these crops. Other important vegetable crop landraces grown were pumpkin, turnip, and watermelon, which were found in 33%, 50%, and 33% of villages assessed, respectively. The cultivation of yellow carrot landraces in half the villages was also unexpected. It appears that yellow carrots are a regional specialty and are often used as a substitute for the more traditional orange carrots that are oftentimes used in couscous. Interestingly, no tomato landraces were identified in any village assessed (Table 3).

Table 3. Vegetable crop landraces currently grown, and those lost over the last 30 years based on an assessment of small, subsistence farmers in six villages in the Souss-Massa region of Morocco in 2014.

Moroccan Souss-Massa Region Village	Current Cultivated Vegetable Crop Landraces	Vegetable Crop Landraces Lost within Last 30 Years
Anou eljdid	carrot (yellow), melon, turnip	carrot, fava bean, melon, pumpkin, watermelon, tomato
Biougra	melon, onion, watermelon	melon, tomato
Massa	melon, onion, watermelon	tomato
Taliouine	carrot (yellow), onion	fava bean, tomato
Targa N' Touchka	carrot (yellow), turnip	melon, watermelon

The latitude (°N) and longitude (°W) of each village assessed: Anou eljdid (30.162045, −9.252229); Biougra (30.220658, −9.372859); Massa (30.004087, −9.636919); Taliouine (30.531266, −7.923879); Targa N' Touchka (29.885070, −9.199597); and Tizi N' Test (30.868223, −8.379130). Vegetable crop landrace estimates presented are based on conversations with small vegetable growers in each village.

3.3. Vegetable Crop Landraces Still Grown and Maintained

In addition to providing unique data regarding landraces that are currently cultivated in the Souss-Massa region of Morocco, the number of landraces that are generally maintained by individual farmers was also determined. Some farmers still maintain numerous landraces of several crops, as the farmer visited in Tizi N' Test (Figure 3) maintained a total of 12 vegetable crop landraces, including two each of pumpkin, carrot, and turnip (Table 3). In comparison, farmers in other villages assessed only cultivated and maintained two to three vegetable crop landraces. The utilization of 12 different landraces requires considerable effort for seed-saving activities, such as planning their field location each year to maintain landrace seed purity (especially on a small farm), as well as seed harvesting, cleaning, and storage. This farmer stated that many younger farmers do not use landraces due to all that is required for their maintenance from year to year, and the use of landraces is becoming a lost tradition.

Figure 3. A Moroccan small landholder farmer near the village of Tizi N' Test (in the Anti-Atlas mountain region of southwestern Morocco) showing off his prized turnip. This turnip is a specimen of the landrace that was collected at this location.

4. Discussion

4.1. Vegetable Landrace Overview and Importance

Landraces of vegetable crops can still be found in many developing countries, but their cultivation is diminishing [4]. The movement toward homogeneity in the world's food supply continues at a rapid rate with no indication of slowing down, and, as a consequence, world food supplies have become more similar in composition, especially important cereal crops [9]. A serious consequence of modern monoculture cropping systems utilizing genetically-uniform hybrids and improved varieties is the significant loss of crop biodiversity through displacement of locally adapted landraces [5]. Plant breeders throughout the world are engaged in developing better and higher yielding varieties of crop plants, with their adoption resulting in at least the partial replacement of the more diverse, genetically-variable, lower yielding, locally-adapted varieties or landraces [10]. Thus, crop genetic uniformity is replacing diversity in many of the world's cropping systems, even with the existing and ongoing genetic diversity management practices of subsistence farmers [11]. Moreover, this crop genetic diversity, which is crucial for most of the world's sustainable agricultural systems that feed most of humanity are being lost at an alarming rate [6]. Genetic diversity is essential to improve crop productivity, to enhance ecosystem functions, and to provide sustainability and adaptability over time [12].

Vegetable crop landraces are often used by small landholders in developing countries and the genetic diversity within these landraces may be the only resource available to allow them to cope with changing environmental conditions to optimize crop production. Landrace populations grown by subsistence farmers may be highly competitive with modern hybrids, especially if they are grown using sustainable minimal-input production practices [13], since landrace populations generally evolved in these types of farming systems [14]. Agricultural biodiversity will also be absolutely essential to cope with the predicted impacts of climate change, not simply as a source of genes, but as the base for more resilient farm ecosystems, which is particularly pertinent in areas where diverse production systems still prevail, most notably marginal agricultural areas [12].

4.2. Significant Genetic Erosion and Loss of Vegetable Landraces

Our assessment indicated that even in remote areas of the Souss-Massa region of Morocco that were visited, including Taliouine, Targa N' Touchka, and Tizi N' Test, a significant loss of vegetable crop landraces has occurred during the last 30 years with approximately 80 to 90% of vegetable crop landraces lost during this time (Table 2). This loss has included a diversity of vegetable crops. The most significant loss observed was tomato, as no landraces of this crop species were found in this region, which indicated that tomato landraces that were grown only a few years ago are now completely lost. Much of this deterioration of tomato crop genetic diversity has primarily resulted from activities of large agribusiness farms who export much of their products to Europe. Western nations in Europe demand a high-quality product for their consumers, with hybrid varieties providing the most disease-resistant plants and best quality fruit. Additionally, some of this higher quality fruit is also marketed in Morocco, which drives the domestic market towards the use of hybrids and away from tomato landraces. Furthermore, the Tomato Yellow Leaf Curl Virus (TYLCV), which is transmitted by whiteflies, is a devastating disease of tomato in Morocco, and has also had an influence on perpetuating the use of hybrid tomato varieties. Tomato crops must either have resistance to this disease or be grown in greenhouses with whitefly exclusion screening to prevent this insect from transmitting the pathogen. This situation has directly led to the widespread use of tomato hybrids with TYLCV resistance, since landraces do not have this resistance.

The genetic deterioration of vegetable crop landraces in developing countries, such as Morocco, is disturbing. This should have been expected because they are being quickly replaced in developing countries by newer hybrids, or even older pure-line inbred varieties, which have superior yields and product quality [4,15]. Moreover, the recent genetic deterioration of all vegetable crops has directly resulted from consumer demand for superior, high quality vegetable products in the marketplace, even in remote locations of the world [4]. Markets have changed within the last few decades in developing countries due to the influx of high-quality vegetable seeds/produce from developed countries that result primarily from improved vegetable crop varieties, and this has definitely had a detrimental influence of the cultivation of local crop varieties. Once consumers become adjusted to the high and consistent quality of these vegetable products, they will not accept inferior produce that oftentimes result from the cultivation of landraces. This most likely directly resulted in the loss of tomato landraces being grown and maintained in the Souss-Massa region of Morocco due to the resulting poor-quality characters, such as the high proportion of cracked or misshapen fruit, poor flavor, and low yield. Tomato landrace populations tend to be high heterogeneous mixtures that often provide many differing phenotypes under field conditions, which relate to their high amounts of genetic diversity. However, large amounts of heterogeneity in a landrace population will most likely make it unattractive to growers for several reasons, including the variation in disease control, multiple plant and fruit sizes at maturity, and inconsistent product quality and yield. The high heterogeneity of fruit and yield characters within landrace populations, limit their use for commercial plantings [2]. Thus, it is essential to identify populations that are consistent in quality and appearance (or similar to pure lines), since most farmers today are only willing to cultivate homogeneous vegetable crop populations.

4.3. Landraces and Food Security

Landraces are low-cost and sustainable, which is important to poor households in marginal environments that have limited monies available, such as those found in the Souss-Massa region of Morocco. The purchase of other sources of seed having much higher associated costs can leave these households vulnerable to chronic food insecurity [3]. The utilization of landrace populations by growers is critical for many reasons, as this allows these important resources to be perpetuated and conserved on-farm for future generations. Crop genetic resources are crucial for the future survival of humanity, and future food security depends upon their conservation [7]. Thus, without some type of on-farm genetic conservation for these crops, there is a bleak outlook for global food security [4].

4.4. Associated Impacts of Green Revolution on Vegetable Landraces

Since the beginning of the Green Revolution in the late 1960s and early 1970s [16], the resulting changes has been significant regarding the development and utilization of improved varieties in most cropping systems besides those high caloric crops that were initially targeted for change, which included maize, rice, and wheat. This situation has contributed significantly to the depletion of genetic diversity for most major crops grown throughout the world [4]. The results of the Green revolution has in a way caused our food systems to fail globally by not providing enough balanced nutrient/vitamin output to meet all the nutritional requirements of every person, especially resource-poor women, infants, and children in developing countries [17]. Moreover, improved crop development activities stemming from this movement have also definitely contributed to the loss of vegetable crop landraces throughout the world. The genetic diversity that exists in crop landraces is not only an economically valuable part of global diversity, but it is also of paramount importance for future world crop production [18]. Plant genetic resources remain fundamental to our efforts to improve world agricultural productivity [10]. Due to the rapid decline of landraces observed during the last half-century in many of the world's cropping systems, the genetic diversity contained within future world crop production systems will be significantly narrowed even more unless some attention is given by political infrastructures/organizations to maintain on-farm landrace conservation.

4.5. Improving Landrace Demand in the Local Marketplace

The Green Revolution resulted in widespread availability of hybrid vegetable seed to farmers throughout the world, which directly related to a general abandonment of vegetable landraces [19]. Although landrace utilization has definitely significantly diminished over the last few decades in Morocco, there are some instances in which specific vegetable crop landraces can still be used to provide a viable source of income. For example, in Morocco, the culture is accustomed to exceptionally sweet melons, and many landraces are still grown that have this specific characteristic. Ananas and Souehla are two types of melons that are still grown as landraces in Morocco due to their traditional consumer appeal. Both melons are very sweet and widely popular in Morocco, and also have a wide adaptability to be grown in the arid, marginal climates of this country. It is widely known that local products can enjoy a market premium since some consumers recognize their link to local culture and tradition, and are willing to pay higher prices for them [20]. Moreover, landraces that have a niche market as a traditional product are more likely to be maintained by growers [21]. Thus, the development of markets focused on local products is a method that could be used to promote landraces and to ensure their on-farm perpetuation. Although the preservation of local traditions and cultural identity, as well as product taste are all important reasons for growers to produce and maintain landraces [21], they must have an economically viable position in the marketplace before growers will make the long-term commitment to continue their on-farm perpetuation.

5. Conclusions

Landraces remain an important part of vegetable production systems to small landholder, subsistence farmers in the Souss-Massa region of Morocco, although their utilization in low-input, sustainable cropping systems is rapidly dwindling. Our assessment indicated that landraces of some vegetable crops, like melons, onions, and turnips, are still widely grown throughout this area of Morocco, and contribute substantially to domestic income from sales at local weekly markets. However, many different vegetable crops are rapidly being lost, and some crops (e.g., tomato) are no longer maintained as landraces. This situation is unfortunate since landrace populations have often evolved under strenuous climatic conditions and are well adapted to changing environmental conditions. The high amount of genetic diversity in these crop populations allows them to adapt to drought, heat, saline soil, of other extreme environmental conditions which is essential for maintaining long-term crop productivity in stressful climates, such as those found in the Souss-Massa region of Morocco. This genetic diversity is important, since it allows landrace crop populations to adapt to local effects of climate change, and to eventually produce improved yields unique to that environment. However, the utilization of heterogeneous landrace populations depends on the farmer's interest and abilities, and many are only willing to cultivate homogenous lines (e.g., pure lines or hybrids) for multiple reasons. In Morocco, vegetable crop landraces are often used by small landholders and the genetic diversity within their landraces may be the only resource available to these farmers that allows them to cope with changing environmental conditions to optimize crop production. Thus, landraces remain critical components of crop production systems for many small farmers in this region of Morocco.

Acknowledgments: The authors wish to thank the U.S. State Department Fulbright Research Scholar Program (through the Moroccan-American Commission for Educational and Cultural Exchange), Southern Illinois University—Carbondale, and the Moroccan National Institute of Agronomic Research, who all provided support for this project.

Author Contributions: Stuart Alan Walters traveled to the assessed sites, interpreted the results, and wrote most of manuscript. Rachid Bouharroud, Abdelaziz Mimouni, and Ahmed Wifaya all made grower contacts in the villages assessed, traveled to various villages, and provided interpretation services. Rachid Bouharroud also spent significant time editing the manuscript and following up with growers to obtain additional required information.

Conflicts of Interest: The authors declare no conflict of interest.

References

1. Camancho Vila, T.C.; Maxted, N.; Scholten, M.A.; Ford-Lloyd, B.V. Defining and identifying crop landraces. *Plant Genet. Resour.* **2005**, *3*, 373–384. [CrossRef]
2. Terzopoulos, P.J.; Walters, S.A.; Bebeli, P.J. Evaluation of Greek tomato landrace populations for heterogeneity of horticultural traits. *Eur. J. Hortic. Sci.* **2009**, *74*, 24–29.
3. Morris, M.L.; Bellon, M.R. Participatory plant breeding research: Opportunities and challenges for the international crop improvement system. *Euphytica* **2004**, *136*, 21–35. [CrossRef]
4. Walters, S.A. Vegetable seed availability and implications for developing countries: A perspective from Morocco. *Outlook Agric.* **2016**, *45*, 18–24. [CrossRef]
5. Sangam, D.; Ceccarelli, S.; Blair, M.W.; Upadhyaya, H.D.; Kumar, A.; Ortiz, R. Landrace germplasm for improving yield and abiotic stress adaptation. *Trends Plant Sci.* **2016**, *21*, 31–42.
6. Esquintas-Alcázar, J. Protecting crop genetic diversity for food security: Political, ethical and technical challenges. *Nat. Rev.* **2005**, *6*, 946–953. [CrossRef] [PubMed]
7. Gepts, P. Plant genetic resources conservation and utilization: The accomplishments and future of a societal insurance policy. *Crop Sci.* **2006**, *46*, 2278–2292. [CrossRef]
8. Brush, St.B.; Meng, E. Farmers' valuation and conservation of crop genetic resources. *Genet. Resour. Crop. Evol.* **1998**, *45*, 139–150. [CrossRef]
9. Khoury, C.K.; Bjorkman, A.D.; Dempewolf, H.; Ramirez-Villegas, J.; Guarino, L.; Jarvis, A.; Rieseberg, L.H.; Struik, P.C. Increasing homogeneity in global food supplies and implications for food security. *Proc. Natl. Acad. Sci. USA* **2014**, *111*, 4001–4006. [CrossRef] [PubMed]
10. Hoisington, D.; Khairallah, M.; Reeves, T.; Ribaut, J.-M.; Skovmand, B.; Taba, S.; Warburton, M. Plant genetic resources: What can they contribute toward increased crop productivity? *Proc. Natl. Acad. Sci. USA* **1999**, *96*, 5937–5943. [CrossRef] [PubMed]
11. Maxted, N.; Guarino, L.; Myer, L.; Chiwona, E.A. Towards a methodology for on-farm conservation of plant genetic resources. *Genet. Resour. Crop Evol.* **2002**, *49*, 31–46. [CrossRef]
12. Frison, E.A.; Cherfas, J.; Hodgkin, T. Agricultural biodiversity is essential for a sustainable improvement in food and nutrition security. *Sustainability* **2011**, *3*, 238–253. [CrossRef]

13. Mavromatis, A.G.; Arvanitoyannis, I.S.; Chatzitheodorou, V.A.; Khan, E.M.; Korkovelos, A.E.; Goulas, C.K. Landraces versus commercial common bean cultivars under organic growing conditions: A comparative study based on agronomic performance and physiochemical traits. *Eur. J. Hortic. Sci.* **2007**, *72*, 214–219.
14. Frankel, O.H.; Brown, A.H.D.; Burdon, J.J. *The Conservation of Plant Biodiversity*; Cambridge University Press: Cambridge, UK, 1995; 299p.
15. Walters, S.A.; Groninger, J.W.; Myers, O. Rebuilding Afghanistan's agricultural economy: Vegetable production in Balkh province. *Outlook Agric.* **2012**, *41*, 7–13. [CrossRef]
16. Evenson, R.E.; Gollin, D. Assessing the impact of the green revolution, 1960 to 2000. *Science* **2003**, *300*, 758–762. [CrossRef] [PubMed]
17. Welch, R.M.; Graham, R.D. A new paradigm for world agriculture: Meeting human needs productive, sustainable, nutritious. *Field Crops Res.* **1999**, *60*, 1 10. [CrossRef]
18. Wood, D.; Lenné, J.W. The conservation of agrobiodiversity on-farm: Questioning the emerging paradigm. *Biodivers. Conserv.* **1997**, *6*, 109–129. [CrossRef]
19. Brush, S.B. The environment and native Andean Agriculture. *Am. Indíg.* **1980**, *40*, 161–172.
20. Negri, V.; Tiranti, B. Effectiveness of in situ and ex situ conservation of crop diversity: What a *Phaseolus vulgaris* L. landrace case study can tell us. *Genetica* **2010**, *138*, 985–998. [CrossRef] [PubMed]
21. Riu-Bosoms, C.; Calvit-Mir, L.; Reyes-García, V. Factors enhancing landrace in situ conservation in home gardens and fields in Vall De Gósol, Catalan Pyrenees, Iberian Peninsula. *J. Ethnobiol.* **2014**, *43*, 175–194. [CrossRef]

agriculture

MDPI

Article

Quality and Nutritional Evaluation of Regina Tomato, a Traditional Long-Storage Landrace of Puglia (Southern Italy)

Massimiliano Renna [1,2], Miriana Durante [3], Maria Gonnella [1,*], Donato Buttaro [1], Massimiliano D'Imperio [1], Giovanni Mita [3] and Francesco Serio [1,*]

[1] Institute of Sciences of Food Production (ISPA), CNR, via Amendola 122/O, 70126 Bari, Italy; massimiliano.renna@uniba.it (M.R.); donato977@gmail.com (D.B.); massimiliano.dimperio@ispa.cnr.it (M.D.)

[2] Department of Agricultural and Environmental Science, University of Bari Aldo Moro, via Amendola 165/A, 70126 Bari, Italy

[3] Institute of Sciences of Food Production (ISPA), CNR, via Lecce-Monteroni, 73100 Lecce, Italy; miriana.durante@ispa.cnr.it (M.D.); giovanni.mita@ispa.cnr.it (G.M.)

* Correspondence: maria.gonnella@ispa.cnr.it (M.G.); francesco.serio@ispa.cnr.it (F.S.); Tel.: +39-080-5929306 (M.G.); +39-080-5929313 (F.S.)

Received: 10 May 2018; Accepted: 11 June 2018; Published: 13 June 2018

Abstract: Regina tomato, a locally cultivated Italian landrace, is listed as an item in the 'List of Traditional Agri-Food Products' of the Italian Department for Agriculture and itemised as 'Slow Food presidium' by the Slow Food Foundation. It is classified as a long-storage tomato since it can be preserved for several months after harvest thanks to its thick and coriaceous skin. Three ecotypes were investigated for main physical and chemical traits both at harvest and after three months of storage. Experimental results indicate that this tomato landrace has a qualitative profile characterized by high concentrations of tocopherols, lycopene and ascorbic acid (maximum 28.6 and 53.7 mg/kg fresh weight, FW, and 0.28 mg/g FW, respectively) even after a long storage time, together with lower average Total Soluble Solids. The initial and post-storage contents of the bioactive compounds changed at a different rate in each ecotype (i.e., in Monopoli Regina tomato the highest content of α-Tocopherol, thereafter reduced to the same level of the other two ecotypes). These results indicate unique and unmistakable features of this long-storage tomato, closely linked to the geographic origin area that include both natural (available technical inputs) and human (specific cultural practices) factors.

Keywords: ecotypes; geographical origin area; HPLC analyses; long storage time; *Solanum lycopersicum* L.; heirloom

1. Introduction

Tomato (*Solanum lycopersicum* L.) is a species of great economic importance, which is widely cultivated all over the world. It offers beneficial effects to human health through its high content in potassium and antioxidants such as ascorbic acid, vitamin A, lycopene and tocopherols [1]. Tomato was introduced from America to Europe at the beginning of the Sixteenth century with greater success especially in the Mediterranean countries [2]. In Italy, this species found a secondary centre of diversification, since several landraces developed in different regions as a consequence of the adaptation to different environmental and cultivation conditions [3]. This allowed the diffusion of different fruit typologies such as flat, angled, ribbed, pear-shaped, heart-shaped, elongated as well as oval/round, cherry and plum forms [2,4].

Puglia, located in the southern part of Italy, can be considered a region placed in the centre of the Mediterranean basin [5]. With a long tradition in vegetable growing, this region is particularly rich in tomato landraces obtained by farmers themselves after repeated simple selection procedures generation after generation. It is important to highlight that a landrace, also called local variety, farmer's variety, folk variety, is a population of a crop characterized by greater or lesser genetic variation, which is however well identifiable and which usually has a local name [6].

Regina is the name of a tomato landrace grown in the coastal saline soils of the central Puglia. This landrace takes its Italian name, Regina (Italian for "Queen"), from its calyx (Figure 1A), which takes the shape of a crown as it grows. It is listed as an item in the 'List of Traditional Agri-Food Products' of the Italian Department for Agriculture, since its processing, preservation and ageing methods are consolidated in time, harmonious for all the region involved, according to traditional rules, for a period not less than 25 years. Moreover, this landrace is registered also as 'Slow Food presidium' by the Slow Food Foundation [7] that aims to protect authenticity and origin of Italian traditional food products by valorising a geographical area and stimulating market opportunities.

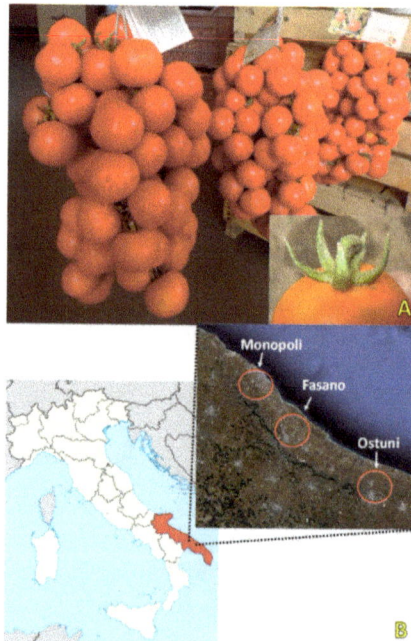

Figure 1. Bunches, so-called ramasole, of Regina tomato with a focus on the fruits calyx (**A**); map of Italy with Puglia region (highlighted in red) and a focus of the three sites (within the circle) where the Regina tomato ecotypes were cultivated and collected (**B**).

The Regina tomato (RT) fruits are small and rounded with a thick skin which improves the resistance to parasites as well as the post-harvest preservation. There is a local tradition of braiding tomatoes in bunches so-called ramasole, by tying the peduncles together with a cotton thread (Figure 1A). The original seed of RT was jealously preserved and reproduced each year by local farmers, who continued to grow it in traditional orchards, even intercropping tomatoes with olive or cotton, as reported by the local history [7]. The hot-arid conditions of the local environment favoured the adaptation of this landrace to the scarcity of the natural resources. So, RT plants are generally grown without irrigation, or with sporadic supplemental irrigation by using brackish water coming from aquifers. This agronomical practice favours the obtaining of tasty tomato fruits very appreciated by local consumers.

RT is locally classified as pomodoro da serbo, in English long-storage tomato, that is a typical southern Italy type of tomato, characterized by small size fruits (i.e., high skin/volume ratio), with thick and coriaceous skin, low water content and rich in bioactive compounds concentrated in the epicarp and mesocarp [4,8,9]. Its fruits are stored in typical hanged crowns for six months [6]. So, a peculiarity of RT is that the ramasole can be preserved for several months ensuring availability of this kind of tomatoes in the winter months. Nevertheless, over the last 20 years, the cultivation of this local variety has gradually diminished due to the advent of greenhouse production that brings cherry tomatoes on the market all the year-long.

A recent and remarkable study defines a thorough nutritional profile of several long storage tomato genotypes retrieved in different area of Southern Italy [10]. Studies are available for other tomato landraces of different geographical areas [1,2,11,12], but literature lacks information on RT, a landrace of an unreported geographical area. Starting from these remarks, the aim of the present study was to evaluate the main physical and chemical traits of this heirloom tomato at harvest and after three mounts of storage, simulating the long storage of the ramasole. The general goal was to assess quality and nutritional traits of RT and promote its diffusion in the market as a regional origin food product recognized by specific marks such as the Protected Designation of Origin (PDO) and the Protected Geographical Indication (PGI).

2. Materials and Methods

2.1. Location and Cropping Details

Three RT ecotypes were open field cultivated on local private farms by using local cultivation practices. Each ecotype was cultivated inside the area where it has been selected: Monopoli (40°55′40.8″ N 17°19′20.0″ E), Fasano (40°51′27.8″ N 17°24′16.3″ E) and Ostuni (40°46′05.7″ N 17°35′43.8″ E) (Figure 1B), since each ecotype is characterized by a specific adaptation to the environmental and the cultivation conditions of the particular area where it has been selected. For this reason, each ecotype has a local name that corresponds to the name of the cultivation site.

Seeds for crop propagation were self-produced and stored on-farm by each farmer. Transplanting was carried out at the four-leaf stage on 15 May for all sites. Plant density was 3.3 plants/m^2 and growing techniques were in line with the agricultural practices of local farmers specialized in growing this type of tomato. Plants were cultivated on silty-sandy soil, typical of the Puglia coastal areas, under the Mediterranean climate, with rain almost absent during the summer season and maximum temperatures sometimes approaching 30–35 °C on the hottest days. So, supplemental irrigation was carried out, only as rescue irrigation after transplanting, for a total water amount of approximately 30 mm applied in one-two times, according to each site needs. Harvesting took place on 5 September for both sites when the proportion of ripe fruit reached about 95%. At harvest tomato samples were refrigerated and then transported to the laboratory to be processed and analysed as described in the following sections.

2.2. Physical Analysis

For each site, three replications of RT fruits were collected for the physical analysis at harvest. So, on twenty fresh fruits for each replication, weight, polar and equatorial diameter, colour parameters and Total Soluble Solids (TSS) were measured.

After measurement of fruit sample fresh weight, the colour parameters, L* (lightness), a* (redness) and b* (yellowness) were measured in triplicate at five random points on the peel surface of the fruits with a colorimeter (CR-400, Konica Minolta, Osaka, Japan) in reflectance mode using the CIE L*a*b* colour scale. Before the measurements, the colorimeter was calibrated with a standard reference with L*, a* and b* values of 97.55, 1.32 and 1.41, respectively. Hue angle ($h° = \tan^{-1}[b*/a*]$) and saturation or chroma (C = $[a^{*2} + b^{*2}]^{1/2}$) were then calculated from the primary L^*, a^* and b^* readings. In addition, a Colour Index was calculated, according to the formula: Colour Index = $2000 \times a^*/L^* \times C$, together with the a^*/b^* ratio [13].

The content of TSS, expressed in °Brix, was measured using a refractometer (model DBR35, XS Instruments, Carpi, MO, Italy) on a liquid extract obtained by homogenizing 100 g of tomato fruits from each replication in a blender (Sterilmimex Lab., International PBI, Milan, Italy), and then filtering the juice.

2.3. Samples Preparation for Chemical Analysis

For chemical analysis RT fruits of each site were divided into two equal portions. One portion was stored at 4 °C for 96 days, while the other portion was analysed as fresh samples. For each treatment three replications were prepared for obtaining a total of 18 samples (3 ecotypes × 2 storage times × 3 replications). One part of each replication was dried in a forced air oven at 105 °C until reaching a constant mass for the determination of the dry weight (DW) content. Results were expressed as g/100 g fresh weight (FW). The other half portion was lyophilized by using a laboratory freeze-dryer (LABCONCO FreeZone® Freeze Dry System, model 7754030, Kansas City, MI, USA) equipped with a stoppering tray (LABCONCO FreeZone® Stoppering Tray Dryer, model 7948030, Kansas City, MI, USA). The freeze-dried matter was ground at 500 µm by using a laboratory mill (Retsch Italia s.r.l., Torre Boldone, BG, Italy) to obtain a homogeneous powder.

2.4. Isoprenoids Analysis

Triplicate aliquots of freeze-dried tomato powder were used to extract isoprenoids (carotenoids and tocopherols) by the method of Sadler et al. [14] modified by Perkins-Veazie et al. [15]. HPLC analyses were carried out using an Agilent 1100 Series HPLC system as described by Fraser et al. [16]. Isoprenoids were separated using a reverse-phase C30 column (5 µm, 250 × 4.6 mm) (YMC Inc., Wilmington, NC, USA) with mobile phases consisting of methanol (A), 0.2% ammonium acetate aqueous solution/methanol (20:80 v/v) (B), and tert-methyl butyl ether (C). The isocratic elution was as follows: 0 min, 95% A and 5% B; 0 to 12 min, 80% A, 5% B, and 15% C; 12 to 42 min, 30% A, 5% B, and 65% C; 42 to 60 min, 30% A, 5% B, and 65% C; 60 to 62 min, 95% A, and 5% B. The column was re-equilibrated for 10 min between runs. The flow rate was 1.0 mL/min, and the column temperature was maintained at 25 °C. The injection volume was 10 µL. Absorbance was registered by diode array at wavelengths of 475 nm for carotenoids and 290 nm for tocopherols. Peaks were identified by comparing their retention times and UV-vis spectra to those of authentic standards. Tocopherols and carotenoid standards were purchased from Cayman chemicals (Ann Arbor, MI, USA) and CaroteNature (Lupsingen, Switzerland), respectively. The limit of detection was 0.4 mAU, tipically in the 2–10 ng/g range per compound. α-Tocopherol (code 10007705) and ß-tocopherol (code 46401-U) were purchased from Cayman chemicals (Ann Arbor, MI, USA) and Sigma Aldrich (Milan, Italy) respectively. Lutein (code 0133), α-carotene, (code 0007), ß-carotene (code 0003) and lycopene (code 0031) were purchased from CaroteNature (Lupsingen, Switzerland).

2.5. Ascorbic Acid Determination

The ascorbic acid content was evaluated using the method described by Ferreira et al. [17] with some modifications. Freeze-dried tomato samples (0.1 g) were extracted by 10 mL of 1% (w/v) metaphosphoric acid followed by shaking for 45 min at room temperature. The extract was centrifuged at 4000× g for 10 min. The supernatant was collected and used for further analysis.

To 1 mL of supernatant, 9 mL of 0.005% 2,6-dichlorophenolindophenol (DCPIP) was added and the absorbance was measured within 30 min at 515 nm against a blank. The content of ascorbic acid was calculated on the basis of the calibration curve of authentic L-ascorbic acid (25–250 µg/mL; $Y = -0.0048x + 1.2708$, $R^2 = 0.9994$).

2.6. Glucose and Fructose Assay, and Sweetness Index

Triplicate aliquots of freeze-dried tomato powder were used to determine glucose and fructose contents, by ionic chromatography (Dionex model DX500; Dionex Corp., Sunnyvale, CA, USA) according to protocols used by Caretto et al. [18]. Results were expressed as mg/g FW.

The sweetness index (SI) was calculated based on content and sweetness properties of individual carbohydrates by multiplying the sweetness coefficient of each sugar (glucose = 1.00 and fructose = 2.30) by the concentration (g/100 g FW) of that sugar in fruits [19].

2.7. Starch Determination

Starch content was determined by using Starch Assay Kit. The samples, 100 mg, were previous washed with aqueous ethanol, 10 mL (80% v/v), incubated at 80–85 °C for 5 min and stirred. After stir the samples were centrifuged for 10 min at $1800\times g$ at room temperature, the supernatants were removed, and the samples were washed with aqueous ethanol (80% v/v) for three different times, in order to remove the D-glucose. The pellets were used for quantification of starch in samples by using Megazyme kit total starch amyloglucosidase/α-amylase (K-TSTA, Megazyme International Ltd. Wicklow, Bray, Ireland). For each treatment, triplicate extractions and analyses were carried out.

2.8. Inorganic Cation Contents

The content of Na^+, K^+, Mg^{2+} and Ca^{2+} was determined by ion exchange chromatography (Dionex DX120, Dionex Corporation, Sunnyvale, CA, USA) with a conductivity detector using an IonPac CG12A guard column and an IonPac CS12A analytical column (Dionex Corporation) as reported by D'Imperio et al. [20]. Results were expressed as mg/g FW.

2.9. Statistical Analysis

A two-way analysis of variance (ANOVA) was performed using the GLM procedure (SAS software, Version 9.1) applying a strip-plot design with storage time and site of cultivation as main factors for carotenoids, tocopherols, ascorbic acid, glucose and fructose content, dry matter and inorganic cations. For other quality traits a one-way ANOVA was performed considering only the ecotype. The separation of means was obtained by the Student–Newman–Keuls (SNK) test.

For a visual analysis of the data, Principal Component Analysis (PCA) (PROC PRINCOMP, SAS Software, Cary, NC, USA) was performed on previously mean-centred and standardised (unit-variance-scaled) data. The data matrix submitted to PCA was made up of six observations (two storage times and three sites of cultivation) and 17 parameters (Na^+, K^+, Mg^{2+} and Ca^{2+}, DM, Glucose, Fructose, Glucose/Fructose ratio, Starch, Sweetness index, Lutein, α-Carotene, β-Carotene, Lycopene, $\beta+\gamma$ tocopherol, α-tocopherol, ascorbic acid). The PCA was applied to obtain an overview of the whole data variability simplified in a few main information. The results of the PCA are shown as biplots (XLStat, Addinsoft, Paris, France) of scores (storage time x ecotype) and loadings (variables).

3. Results and Discussion

3.1. Physical Traits at Harvest

RT fruits showed an average weight of about 22 g without any differences between the three ecotypes. At the same time, the average polar and equatorial diameter were, respectively, 29.9 and 34.5 mm without any differences between the three ecotypes (Table 1). On the other hand, TSS in Ostuni fruits were 9% and 19% higher than Monopoli and Fasano, respectively. These results suggest that different ecotypes did not affect form and size of the RT fruits, whereas TSS was influenced by different ecotypes as well as by environmental conditions. Moreover, these results are in according to some Authors [8,21] who described fruits of long-storage tomatoes with an average weight of about 10–25 g and a form round or ellipsoid. It is interesting to highlight that form and size of RT are similar to "cherry tomato" fruits, considering that this type of tomato generally shows a diameter between 15 and 35 mm and an average weight between 10 and 30 g [21]. On the other hand, RT showed a lower TSS content respect to some cherry tomato types [21] as well as some landrace of Italian long-storage tomatoes such as "Montallegro" and "Filicudi" [22]). At the same time the TSS content of RT fruits results are similar to other Italian long-storage tomatoes such as "Albicocca di Favignana" and "Giallo piccolo a punta" [23] and other types of tomato including landraces and hybrids [2].

Regarding colour analysis, Monopoli fruits showed the highest lightness (L^*), yellowness (b^*), Hue angle ($h°$) and Chroma (C) value, while no difference between the ecotypes was found as regards redness (a^*). No differences were found for C between Ostuni fruits and those from Monopoli and Fasano. These results suggest that the colour of Fasano and Ostuni fruits may be perceived as more "red" than that of Monopoli fruits, according to Serio et al. [21] who described fruits of pomodoro da serbo with skin colour ranging from yellow to red. The colour index and a^*/b^* ratio confirm this perception. Higher values of these two indices are assigned to redder and more mature tomato fruits with a prevalence of red colour [13,24]. In the case of RT fruits, lower values of L^* and C in Fasano and Ostuni fruits influence the colour index values and differences in b^* values are determinant for the values of a^*/b^* ratio (Table 1). Tomatoes are commonly selected by consumers on the basis of appearance, and colour is one of the most important quality factors that affect tomato appearance [25]. Tomato fruits are available in different external colours including pink, yellow, orange and red, as the result of both flesh and skin colours. Thus, a pink tomato may be due to colourless skin and red flesh, while an orange tomato may be due to yellow skin and red flesh [25]. RT fruits as well as other types of Italian long-storage tomatoes are characterized by the presence of a thick and coriaceous skin. Therefore, it is possible that the external colour of the RT fruits is strongly affected by their skin colour.

3.2. Dry Matter and Inorganic Cation Contents

The highest and the lowest values of dry matter were found in RT fruits after 96 days of storage, respectively, from Ostuni and Monopoli (Table 2). In Fasano fruits, the storage did not affect the dry matter content (7.18 g/100 g FW on average). The increase of dry matter in Ostuni and Monopoli fruits can be due to water loss by transpiration and respiration during long-term storage [26]. Nevertheless, it is important to highlight that this increase of dry matter can be evaluated as marginal considering the long-term of storage. This is due to the presence of a coriaceous skin on pomodoro da serbo fruits, which strongly reduces the water evaporation and, therefore, favours a long storage period [21]. Baldina et al. [2] report a range of dry matter between 4.76 and 5.20 g/100 g FW on fifteen Italian tomato landraces subjected to characterization. In a study aimed to evaluate the effects of watering regime on the quality of typical long-storage cherry tomatoes, Barbagallo et al. [23] report an average dry matter content between 5.92 and 7.34 g/100 g FW, corresponding to watering regimes of 100% and 50%, referring to the amount of water required to integrate plant evapotranspiration during growth. A high dry matter content in tomato fruits is considered as a good quality trait, and it is well known that a moderate water and/or salinity stress can increase dry matter in tomato fruits [23,27]. The results of the present study suggest that, independent of the ecotype, RT fruits showed a higher dry matter content compared to other types of long-storage tomatoes, highlighting an important good quality trait for the RT landrace. As regards the inorganic cation contents, only in the case of Mg^{2+} a significant difference for the interaction between sites and storage time was found. The storage did not affect the Mg^{2+} content in Monopoli and Fasano fruits, whereas it caused an increase of about 18% in Ostuni. This is in agreement with the increase of the dry matter in Ostuni fruits as previously reported (Table 2). Magnesium is an important element for the human body: it is a cofactor for more than 300 enzymes and also has functions that affect nerve conduction [28]. On average, the Mg^{2+} content in RT fruits results are similar to what was reported by Chapagain and Wiesman [29] and by the United States Department of Agriculture for red tomatoes [30]. At the same time, RT fruits showed a higher and lower Mg^{2+} content as compared to orange and yellow tomatoes, respectively [30].

Table 1. Fruit weight, morphological traits, total soluble solids (TSS) and colour parameters of the Regina tomatoes at harvest.

Ecotype	Value	Fruit Weight (g)	Fruit Diameter (mm)		TSS (°Brix)	L*	a*	b*	h°	C	Colour Index	a*/b*
			Pol	Equ								
Monopoli	Mean	21.5	30.0	34.8	7.0 a	47.0 a	32.2	41.0 a	51.7 a	52.3 a	26.5 b	0.8 b
	SD	3.7	2.1	1.9	0.3	2.7	3.0	4.4	5.2	2.3	4.5	0.2
Fasano	Mean	20.1	29.4	34.0	6.4 c	40.0 b	32.3	30.3 b	43.1 b	44.4 b	36.6 a	1.1 b
	SD	3.3	1.5	2.1	0.1	2.8	4.3	4.6	1.9	6.1	3.2	0.1
Ostuni	Mean	25.0	30.2	34.8	7.6 a	38.1 b	34.4	32.8 b	43.6 b	47.5 ab	38.0 a	1.1 b
	SD	6.3	2.6	4.4	0.3	1.6	2.8	2.6	1.5	3.6	1.9	0.1
Significance [1]		ns	ns	ns	**	***	ns	**	**	*	***	**

[1] Significance ns = not significant; *, ** and *** significant for $p \leq 0.05$, 0.01 and 0.001, respectively. Different letters indicate statistically significant differences at $p = 0.05$.

Table 2. Dry matter, inorganic cations, glucose and fructose content, glucose/fructose ratio, starch and sweetness index of the Regina tomatoes at harvest and after 96 days of storage.

Ecotype	Days of Storage	Value	Dry Matter (g 100 g⁻¹ FW)	Na⁺	K⁺	Ca²⁺	Mg²⁺ (mg·kg⁻¹ FW)	Glucose	Fructose (g·kg⁻¹ FW)	Glu/Fru Ratio	Sweetness Index	Starch (mg·100 g⁻¹ FW)
Monopoli	0	Mean	6.24 d	164.6	2546	83.3	105.1 b	5.9	9.0	0.66	2.7	3.5
		SD	0.18	11.4	14	22.7	3.7	1.6	2.3	0.02	0.7	0.3
	96	Mean	6.80 c	178.9	2448	86.6	108.8 ab	5.7	8.6	0.66	2.6	1.9
		SD	0.18	26.9	114	17.7	14.2	0.5	0.7	0.01	0.2	0.1
Fasano	0	Mean	7.09 b	117.7	2838	74.1	115.5 a	9.3	13.0	0.71	3.9	4.3
		SD	0.06	15.2	89	20.8	9.0	0.2	0.3	0.01	0.1	0.5
	96	Mean	7.28 b	132.0	2909	91.9	116.0 a	6.3	8.0	0.79	2.5	2.9
		SD	0.07	5.0	92	0.9	0.4	0.5	0.7	0.01	0.2	0.1
Ostuni	0	Mean	7.07 b	119.3	2759	108.2	100.5 b	6.9	10.8	0.64	3.2	3.4
		SD	0.06	6.4	173	12.4	10.1	1.1	1.5	0.01	0.4	0.5
	96	Mean	7.66 a	150.6	2966	163.6	118.3 a	5.4	8.7	0.62	2.5	2.2
		SD	0.23	18.5	74	45.8	4.1	0.9	1.2	0.02	0.4	0.1
Significance [1]												
Ecotype (E)			**	*	**	*	*	ns	ns	**	ns	*
Day of storage (E)			ns	*	ns	ns	*	*	*	ns	*	**
E × D			**	ns	ns	ns	*	ns	ns	ns	ns	ns

[1] Significance: ns = not significant; *, ** significant for $p \leq 0.05$ and 0.01, respectively. Different letters indicate statistically significant differences at $p = 0.05$.

Monopoli fruits showed the highest value of Na⁺ and the lowest of K⁺ (Figure 2A,B). As regards Ca²⁺, Ostuni fruits showed the highest value (Figure 2C), 62% higher than in Monopoli and Fasano. The Na⁺ content in Monopoli fruits was 32% higher than in Fasano and Ostuni, whereas the K⁺ content in Fasano and Ostuni fruits was 15% higher than in Monopoli. This is in agreement to values regarding the inverse relationship between Na⁺ and K⁺ reported by some authors [27,31]. The highest Na⁺ content in Monopoli fruits is due both to the higher electrical conductivity (EC) of the soil and to the use of brackish water for irrigation. This is due to the proximity of Monopoli fields to the marine coast (Figure 1B), which gives irrigation water extracted from underground (at a depth of 15–20 m), resulting in typically brackish water (EC ranging from 4 to 6 dS/m).

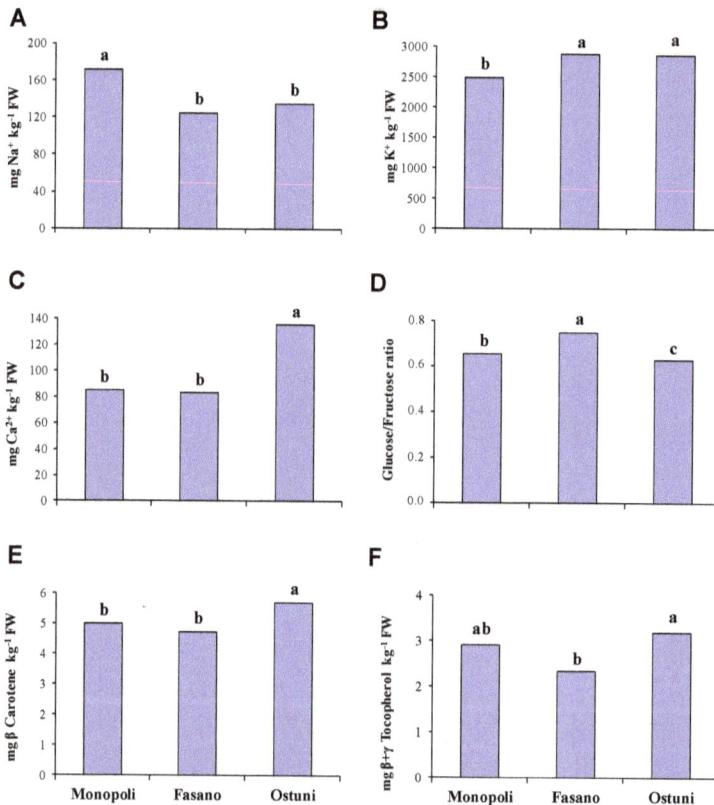

Figure 2. Different Regina tomato traits depending on the ecotype: content of Na⁺ (**A**), K⁺ (**B**), Ca²⁺ (**C**); glucose/fructose ratio (**D**); β-Carotene content (**E**); β + γ Tocopherol content (**F**). The same letters indicate that mean values are not significantly different (*p* = 0.05).

Sodium plays a vital role in the regulation of fluid balance, blood pressure and transmembrane gradients. However, with the aim to prevent heart diseases attributable to high blood pressure, the European Food Safety Authority recommends a maximum intake of 1500 mg of Na⁺ per day [32]. The results of the present study showed that, independently on the cultivation site, RT fruits may be considered as very low contributors to the daily supply of Na⁺, since also in the case of RT from Monopoli, 100 g of fruits supply only about the 1% of the daily intake.

3.3. Sugars, Starch Assay and Sweetness Index

For both sugars and starch, no significant differences were found for the interaction between ecotypes and storage time (Table 2). The results regarding sugar content showed no significant differences between the three ecotypes with the exception of the glucose/fructose ratio (Table 2). At harvest, the average contents of glucose and fructose in RT fruits were, respectively, 7.35 and 10.90 g/kg FW with a reduction of about 21% and 23% after 96 days of storage (Table 2). Regarding starch, the average content at harvest was 3.73 mg/100 g FW with a reduction of about 37% after 96 days of storage (Table 2). The highest and the lowest glucose/fructose ratio was observed in Fasano and Ostuni fruits, respectively (Figure 2D). The average sweetness index (SI) at harvest was of 3.24, while in RT fruits after a 96-day storage period it was 2.52 (Table 2). These results suggest that the content of reducing sugars (the main tomato carbohydrates) in RT fruits was not influenced by the ecotype, whereas it was affected by the storage time. According to Kader [33] the decrease of glucose and fructose in RT fruits can be due to the respiration process during the long-term storage of fruits. At the same time, the starch content too decreased during long-term storage due to the respiration process of the fruits, but at first converted to the monosaccharides glucose and fructose. It is interesting that in Monopoli fruits, marked starch decrease during long-storage did not correspond to a decrease in reducing sugars (quite stable during storage), while in Fasano fruits a drastic reduction of both starch and monocarbohydrate sugars was observed, even without a significant statistical difference in interaction values (Table 2). In addition, it is interesting to highlight that RT showed a lower average glucose and fructose content compared to other types tomatoes. Effectively, in a study aimed to evaluate six genotypes of long-storage tomatoes, Barbagallo et al. [23] report for glucose a range between 14.4 and 21.5 g/kg FW and for fructose a range between 14.7 and 22.0 g/kg FW. Moreover, concerning other types of tomatoes such as "cherry tomato", some Authors [23,31] report, respectively, ranges between 15.9 and 25.0 g/kg FW for glucose and between 16.4 and 27.0 g/kg FW for fructose, in the cultivar Naomi (a diffused commercial variety). As a consequence, a serving size of RT fruits supplies less than the 50% of the glucose and fructose with respect to other landrace and commercial variety of tomatoes. Thus, RT fruits would be preferred with respect to other types of tomatoes in order to reduce the glycemic load of the meals and, therefore, for reducing elevation in blood insulin associated to an increased risk of type 2 diabetes and coronary heart disease [34]. At the same time, it is important to consider that the lower content of sugars in RT fruits also translates into an SI up to four-fold lower compared to other tomato genotypes [23,31]. Therefore, since the SI is one of the common measures of acceptability of horticultural produce and considering the desire of consumers for sweet tomatoes [35], it is possible that the lower SI of the RT fruits may affect consumers' acceptability.

3.4. Isoprenoids Content

RT fruits showed an average lutein content of 5.27 mg/kg FW without any differences between the three ecotypes, the storage time and their interaction (Table 3). The mean content of α- and β-Carotene at harvest was, respectively, 0.6 and 5.7 mg/kg FW with a reduction of about 17% and 18% after 96 days of storage (Table 3); no significant differences were found for the interaction between ecotypes and storage time (Table 3). On the other hand, Ostuni fruits showed a β-Carotene content 17% higher compared with Monopoli and Fasano (Figure 2E). These results are in agreement with those of other authors [10,21,36] who reported as lycopene is the main tomato carotenoid, while α- and β-Carotene and lutein are present in low quantities.

Agriculture **2018**, *8*, 83

Table 3. Carotenoids, tocopherols and ascorbic acid content of the Regina tomatoes at harvest and after 96 day of storage.

Ecotype	Days of Storage	Value	Lutein	α-Carotene	β-Carotene	Lycopene	α Tocopherol	β + γ Tocopherol	Ascorbic Acid
			mg·kg⁻¹ FW						mg·g⁻¹ FW
Monopoli	0	Mean	5.9	0.61	5.4	53.7 a	25.2 a	3.4	0.24 b
		SD	0.2	0.03	0.4	2.8	1.0	0.2	0.03
	96	Mean	5.1	0.47	4.7	37.1 b	19.0 b	2.4	0.19 b
		SD	0.3	0.10	0.4	1.5	2.0	0.4	0.01
Fasano	0	Mean	5.4	0.60	5.4	41.2 b	20.1 b	2.5	0.28 a
		SD	0.3	0.03	0.1	1.8	0.7	0.1	0.02
	96	Mean	4.9	0.48	4.1	37.4 b	19.3 b	2.2	0.09 d
		SD	0.4	0.08	0.1	2.3	0.5	0.1	0.01
Ostuni	0	Mean	5.2	0.58	6.2	41.3 b	21.5 b	3.5	0.20 b
		SD	0.4	0.03	0.3	0.4	1.1	0.6	0.02
	96	Mean	5.1	0.49	5.2	27.3 c	20.3 b	2.8	0.14 c
		SD	0.4	0.09	0.3	2.4	1.7	0.1	0.01
Significance [1]									
Ecotype (E)			ns	ns	*	**	*	*	*
Day of storage (D)			ns	*	**	**	**	**	***
E × D			ns	ns	ns	*	**	ns	**

[1] Significance: ns = not significant; *, ** and *** significant for $p \leq 0.05$, 0.01 and 0.001, respectively. Different letters indicate statistically significant differences at $p = 0.05$.

As regards lycopene, the highest value was found in Monopoli at harvest, with a reduction of 31% after 96 days of storage. The storage did not affect the lycopene content in Fasano, while it caused a 34% reduction in Ostuni (Table 3). Lycopene is responsible for the red colour of tomatoes and is important for human health due to its antioxidant activity. Moreover, dietary intake of tomato lycopene has been shown to be associated with a decreased risk of chronic diseases, such as cancer and cardiovascular disease [21]. The lycopene content of fresh tomato fruits depends on several factors including genotype, environmental growth conditions, cultivation management as well as ripening and phenological stage [21]. Lenucci et al. [37] and Ilahy et al. [38] reported a content of lycopene ranging from 43 to 120 mg/kg FW in ordinary tomato cultivars and 175 to 253 mg/kg FW in high pigment tomato hybrids, grown simultaneously in an open-field of Southern Italy. The results of the present study show that RT fruits can be considered a good source of lycopene even after long-term storage. Indeed, especially for the Monopoli and Fasano fruits, the lycopene content after 3 months of storage was similar to those reported for the cherry tomato cultivar Rubino top (43 mg/kg FW) [37] and the long storage tomato cultivars Licata (34 mg/kg FW), Mezzocachi Montallegro (38 mg/kg FW) and Salina 3 (40 mg/kg FW) [10], all grown in Italy. Some other long storage genotypes showed higher lycopene content [10]. As regards the relationship between colour and lycopene content, it is interesting to note that in some cases, deep red varieties of tomatoes (hybrids so-called "high pigment") can contain more than 180 mg/kg FW [21,37], while some yellow varieties can contain only about 5 mg/kg FW [39]. Thus, several Authors have reported the correlation between colour indexes of skin tomatoes and lycopene content, also suggesting equations for calculating the lycopene content based on the skin colour readings [24]. According to these authors, the redder the skin colour, the higher the lycopene content in tomato fruits. In the present study the highest lycopene content was found in Monopoli fruits at harvest. Nevertheless, as previously reported, Fasano and Ostuni fruits may be perceived as more "red" than those from Monopoli (Table 1). These results suggest that the external colour of RT fruits is strongly affected by their main morphological traits, especially thick and coriaceus skin, and suggest that for long-storage tomatoes, it may be difficult to find a correlation between skin colour indexes and lycopene content.

3.5. Tocopherols and Ascorbic Acid Content

Tomato fruit is an important source of vitamin E which consists of four tocopherols (α, β, γ and δ) and four tocotrienols (α, β, γ and δ) and their amount significantly depends on many biotic and abiotic factors. Furthermore, tocotrienols and tocopherols content and composition differ greatly in different plant tissues [21,40]. α-Tocopherol (α-T) and β, γ T-forms (which, in our system, co-migrated as a single peak) were the tocopherol forms detected in RT fruits. α-T shows the highest biological activity compared to the other forms and, in this study, at any rate, α-T content was even more abundant than the β, γ T-forms (Table 3). The highest value of α-T was found in Monopoli fruits at harvest, with a reduction of about 25% after 96 days of storage; for Fasano and Ostuni the storage did not affect the α-T content (Table 3). The content of β, γ T-forms of the Ostuni fruits was higher than from Fasano, while no differences were found between Monopoli fruits and all the other ones (Figure 2F). The mean content of $\beta + \gamma$ T-forms at harvest was 3.1 mg/kg FW with a reduction of about 19% after 96 days of storage (Table 3). No significant difference was found for the interaction between ecotypes and storage time (Table 3).

In agreement with what was reported by some authors [18,28], the results of the present study highlight that the RT showed a higher tocopherols content with respect to F1 hybrid of tomatoes grown by soilless systems. This could be due to the higher seed content in tomato landraces fruits. In fact, it is known that tomato seeds are a source of tocopherols as reported by some authors [41,42].

Highest and lowest ascorbic acid content were found in Fasano fruits, respectively, at harvest and after 96 days of storage, with a reduction of about 68%. The storage did not affect the ascorbic acid content in Monopoli fruits, while it caused a reduction of 30% in Fasano and Ostuni fruits (Table 3). These results are in agreement with Soto-Zamora et al. [43] who found a decrease of ascorbic acid in tomatoes stored at 4 °C. Ascorbic acid is an essential compound for the human health due to its numerous functions such as prevention of scurvy and maintenance of healthy skin, gums and blood vessels and also because it can reduce the risk of arteriosclerosis, cardiovascular disease and some forms of cancer [44]. The ascorbic acid of fresh tomato fruits is usually about 0.23 mg/g FW [20] depending on several factors including genotype, environmental grown conditions and cultivation management [21]. In a study aimed to evaluate 12 genotypes tomatoes, George et al. [36] report an average ascorbic acid content of about 0.30 mg/g FW in cherry tomatoes and a content of about 0.13 mg/g FW in other genotypes. Thus, the results of the present study suggest that RT fruits can be considered a good source of ascorbic acid even after long-term storage, especially for the Monopoli fruits. On the other hand, the mean value of ascorbic acid found in 28 long storage tomato genotypes was considerably higher (0.63 mg/g FW) [10].

3.6. Principal Component Analysis

From PCA results, we can see that PC1 (43.2%) and PC2 (28.0%) explain overall 71.1% of the data variability (Figure 3). PCA allows us to separate distinctly the two storage times along PC1, since data of all ecotypes at storage 0 were positively correlated to PC1, while those at storage 96 were negatively correlated to the same PC. It is also clear that there is a positive correlation to PC1 of all the organic nutritional parameters which were strongly linked to the fresh status of fruits. On the other hand, DM and inorganic cations, not modified during storage, were strictly correlated to the stored material. PC2 helps to clearly separate Fasano fruits from those of Monopoli, both at the harvest time. The first are characterized by high soluble sugars, starch and a sweetness index, isolated in the same PCA quadrant, while Monopoli, at harvest, was distinguished by high tocopherol content as well as low sugar content (Figure 3). Other considerations are related to some observations such as Ostuni after storage, correlated to high Ca^{2+} and Na^+ content, and Fasano after storage negatively correlated to PC1 due to its low ascorbic acid and α-Carotene content (Figure 3).

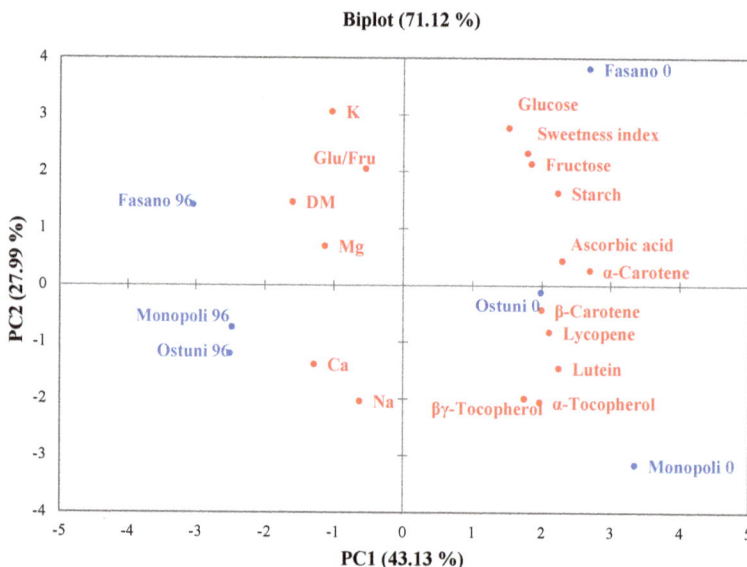

Figure 3. PCA biplot (PC1 vs. PC2) describing the variation of some properties of the three Regina Tomato ecotypes, analysed at harvest (Ecotype 0) and after 96 days of storage (Ecotype 96).

4. Conclusions

Italy is the European country with the largest number of regional origin food products recognized by specific marks such as Protected Designation of Origin and the Protected Geographical Indication. They are assigned by the European Union to those food products that have unique and unmistakable features that depend exclusively, or mainly, on the area where they are produced. In this study, the quality and nutritional assessment of the RT was carried out for highlighting the specific traits of this traditional Italian food product closely linked to the geographic area of origin. Experimental results indicate that this tomato landrace has a qualitative profile that is characterized by high concentrations of tocopherols and a lycopene content similar to values found on average in 28 Italian long storage genotypes. The ascorbic acid, though lower than values found in other Italian tomato landraces, remained quite high even after a long storage time. This profile combined with a lower average glucose and fructose content compared to other types of tomatoes may account for the high nutritional quality of this landrace, especially for people with specific dietary requirements. The results of the present study, enriched by the lacking analysis of the polyphenol content, may be used as a tool for obtaining the PGI or PDO mark, since these results show unique and unmistakable features of this long-storage tomato that depend exclusively on the geographic origin area including both the natural (e.g., climate, environment) and human (e.g., traditional production techniques) factors. Finally, possible next goals may be directed toward non-destructive analytical methods for the rapid determination of nutrients compounds on intact fruits of RT during the long time of storage.

Author Contributions: Conceptualization, M.R. and F.S.; Data curation, M.R., M.D. (Miriana Durante), M.D. (Massimiliano D'Imperio) and F.S.; Formal analysis, M.G.; Funding acquisition, M.R. and F.S.; Investigation, M.R., D.B. and F.S.; Project administration, M.G. and F.S.; Resources, M.R., M.D. (Miriana Durante), M.D. (Massimiliano D'Imperio) and G.M.; Supervision, M.R. and F.S.; Validation, M.R., M.D. (Miriana Durante), M.G., G.M. and F.S.; Visualization, M.R., M.G. and F.S.; Writing—original draft, M.R. and M.G.; Writing—review and editing, M.R., M.D. (Miriana Durante), M.G., D.B., M.D. (Massimiliano D'Imperio), G.M. and F.S.

Funding: This work was supported by Regione Puglia Administration under Rural Development Program 2014–2020—project 'Biodiversity of vegetable crops in Puglia (BiodiverSO)' and "Intervento cofinanziato dal Fondo di Sviluppo e Coesione 2007–2013—APQ Ricerca Regione Puglia—Programma regionale a sostegno della specializzazione intelligente e della sostenibilità sociale ed ambientale FutureInResearch"—project 'Innovazioni di prodotto e di processo per la valorizzazione della Biodiversità Orticola pugliese (InnoBiOrt)'.

Acknowledgments: The authors thank Nicola Gentile and Piero Moretti (Cooperativa Progresso Agricolo) for providing technical assistance during the experiment.

Conflicts of Interest: The authors declare no conflict of interest.

References

1. Naranjo, R.D.D.P.; Otaiza, S.; Saragusti, A.C.; Baroni, V.; Carranza, A.D.V.; Peralta, I.E.; Valle, E.M.; Carrari, F.; Asis, R. Hydrophilic antioxidants from Andean tomato landraces assessed by their bioactivities in vitro and in vivo. *Food Chem.* **2016**, *206*, 146–155. [CrossRef] [PubMed]

2. Baldina, S.; Picarella, M.E.; Troise, A.D.; Pucci, A.; Ruggieri, V.; Ferracane, R.; Barone, A.; Fogliano, V.; Mazzucato, A. Metabolite profiling of Italian tomato landraces with different fruit types. *Front. Plant Sci.* **2016**, *7*, 664. [CrossRef] [PubMed]

3. Mazzucato, A.; Ficcadenti, N.; Caioni, M.; Mosconi, P.; Piccinini, E.; Reddy Sanampudi, V.R.; Sestili, S.; Ferrari, V. Genetic diversity and distinctiveness in tomato (*Solanum lycopersicum* L.) landraces: The Italian case study of "A pera Abruzzese". *Sci. Hortic.* **2010**, *125*, 55–62. [CrossRef]

4. Muratore, G.; Licciardello, F.; Maccarone, E. Evaluation of the chemical quality of a new type of small-sized tomato cultivar, the plum tomato (*Lycopersicon lycopersicum*). *Ital. J. Food Sci.* **2005**, *17*, 75–81.

5. Renna, M.; Rinaldi, V.A.; Gonnella, M. The Mediterranean Diet between traditional foods and human health: The culinary example of Puglia (Southern Italy). *Int. J. Gastron. Food Sci.* **2015**, *2*, 63–71. [CrossRef]

6. Elia, A.; Santamaria, P. Biodiversity in vegetable crops, a heritage to save: The case of Puglia region. *Ital. J. Agron.* **2013**, *8*, 4. [CrossRef]

7. Slow Food Torre Canne Regina Tomato—Presìdi Slow Food—Slow Food Foundation. Available online: https://www.fondazioneslowfood.com/en/slow-food-presidia/torre-canne-regina-tomato/ (accessed on 3 May 2018).

8. Siracusa, L.; Patanè, C.; Avola, G.; Ruberto, G. Polyphenols as chemotaxonomic markers in Italian "long-storage" tomato genotypes. *J. Agric. Food Chem.* **2011**, *60*, 309–314. [CrossRef] [PubMed]

9. Patanè, C.; Pellegrino, A.; Di Silvestro, I. Effects of calcium carbonate application on physiology, yield and quality of field-grown tomatoes in a semi-arid Mediterranean climate. *Crop Pasture Sci.* **2018**, *69*, 411–418. [CrossRef]

10. Siracusa, L.; Patanè, C.; Rizzo, V.; Cosentino, S.L.; Ruberto, G. Targeted secondary metabolic and physico-chemical traits analysis to assess genetic variability within a germplasm collection of "long storage" tomatoes. *Food Chem.* **2018**, *244*, 275–283. [CrossRef] [PubMed]

11. Naranjo, R.D.D.P.; Otaiza, S.; Saragusti, A.C.; Baroni, V.; Carranza, A.V.; Peralta, I.E.; Valle, E.M.; Carrari, F.; Asis, R. Data on polyphenols and biological activity analyses of an Andean tomato collection and their relationships with tomato traits and geographical origin. *Data Brief* **2016**, *7*, 1258–1268. [CrossRef] [PubMed]

12. Figàs, M.R.; Prohens, J.; Raigón, M.D.; Fita, A.; García-Martínez, M.D.; Casanova, C.; Borràs, D.; Plazas, M.; Andújar, I.; Soler, S. Characterization of composition traits related to organoleptic and functional quality for the differentiation, selection and enhancement of local varieties of tomato from different cultivar groups. *Food Chem.* **2015**, *187*, 517–524. [CrossRef] [PubMed]

13. López Camelo, A.F.; Gómez, P.A. Comparison of color indexes for tomato ripening. *Hortic. Bras.* **2004**, *22*, 534–537. [CrossRef]

14. Sadler, G.; Davis, J.; Dezman, D. Rapid extraction of lycopene and β-carotene from reconstituted tomato paste and pink grapefruit homogenates. *J. Food Sci.* **1990**, *55*, 1460–1461. [CrossRef]

15. Perkins-Veazie, P.; Collins, J.K.; Pair, S.D.; Roberts, W. Lycopene content differs among red-fleshed watermelon cultivars. *J. Sci. Food Agric.* **2001**, *81*, 983–987. [CrossRef]

16. Fraser, P.D.; Pinto, M.E.S.; Holloway, D.E.; Bramley, P.M. Application of high-performance liquid chromatography with photodiode array detection to the metabolic profiling of plant isoprenoids. *Plant J.* **2008**, *24*, 551–558. [CrossRef]

17. Ferreira, I.C.F.R.; Aires, E.; Barreira, J.C.M.; Estevinho, L.M. Antioxidant activity of Portuguese honey samples: Different contributions of the entire honey and phenolic extract. *Food Chem.* **2009**, *114*, 1438–1443. [CrossRef]

18. Caretto, S.; Parente, A.; Serio, F.; Santamaria, P. Influence of potassium and genotype on vitamin E content and reducing sugar of tomato fruits. *HortScience* **2008**, *43*, 2048–2051.

19. Crespo, P.; Bordonaba, J.G.; Terry, L.A.; Carlen, C. Characterisation of major taste and health-related compounds of four strawberry genotypes grown at different Swiss production sites. *Food Chem.* **2010**, *122*, 16–24. [CrossRef]

20. D'Imperio, M.; Renna, M.; Cardinali, A.; Buttaro, D.; Serio, F.; Santamaria, P. Calcium biofortification and bioaccessibility in soilless "baby leaf" vegetable production. *Food Chem.* **2016**, *213*, 149–156. [CrossRef] [PubMed]

21. Serio, F.; Ayala, O.; Bonasia, A.; Santamaria, P. Antioxidant properties and health benefits of tomato. In *Progress in Medicinal Plants—Search for Natural Drugs*; Studium Press LLC: Houston, TX, USA, 2006; Volume 13, pp. 159–179, ISBN 9780976184959.

22. Siracusa, L.; Avola, G.; Patanè, C.; Riggi, E.; Ruberto, G. Re-evaluation of traditional Mediterranean foods. The local landraces of "Cipolla di Giarratana" (*Allium cepa* L.) and long-storage tomato (*Lycopersicon esculentum* L.): Quality traits and polyphenol content. *J. Sci. Food Agric.* **2013**, *93*, 3512–3519. [CrossRef] [PubMed]

23. Barbagallo, R.N.; Chisari, M.; Branca, F.; Spagna, G. Pectin methylesterase, polyphenol oxidase and physicochemical properties of typical long-storage cherry tomatoes cultivated under water stress regime. *J. Sci. Food Agric.* **2008**, *88*, 389–396. [CrossRef]

24. Arias, R.; Lee, T.C.; Logendra, L.; Janes, H. Correlation of lycopene measured by HPLC with the L*, a*, b* color readings of a hydroponic tomato and the relationship of maturity with color and lycopene content. *J. Agric. Food Chem.* **2000**, *48*, 1697–1702. [CrossRef] [PubMed]

25. Yahia, E.M.; Brecht, J. Tomatoes. In *Crop Post-Harvest: Science and Technology: Perishables*; Rees, D., Farrell, G.O.J., Eds.; Wiley-Blackwell: Oxford, UK, 2012; pp. 5–23, ISBN 0632057254.

26. Elia, A.; Santamaria, P.; Colelli, G.; De Boni, A.; Ventrella, D.; Serio, F. Produzione, qualità e attitudine alla conservazione di popolazioni di pomodoro da serbo. Risultati preliminari. In *Atti Giornate Scientifiche SOI*; Società Orticola Italiana: Sesto Fiorentino, Italy, 1992; pp. 406–407.
27. Serio, F.; De Gara, L.; Caretto, S.; Leo, L.; Santamaria, P. Influence of an increased NaCl concentration on yield and quality of cherry tomato grown in posidonia (*Posidonia oceanica* L. Delile). *J. Sci. Food Agric.* **2004**, *84*, 1885–1890. [CrossRef]
28. Nielsen, F.H. Trace mineral deficiencies. In *Handbook of Nutrition and Food*; CRC Press, Taylor & Francis: Boca Raton, FL, USA, 2008; pp. 159–176. ISBN 9781420008890.
29. Chapagain, B.P.; Wiesman, Z. Effect of potassium magnesium chloride in the fertigation solution as partial source of potassium on growth, yield and quality of greenhouse tomato. *Sci. Hortic.* **2004**, *99*, 279–288. [CrossRef]
30. USDA Food Composition Databases Show Foods—Tomatoes, Red, Ripe, Raw, Year Round Average. Available online: https://ndb.nal.usda.gov/ndb/foods/show/302115?manu=&fgcd=&ds=SR&q=Tomatoes,red,ripe,raw,yearroundaverage (accessed on 3 May 2018).
31. Elia, A.; Serio, F.; Parente, A.; Santamaria, P.; Ruiz Rodriguez, G. Electrical conductivity of nutrient solution, plant growth and fruit quality of soilless grown tomato. *Acta Hortic.* **2001**, 503–508. [CrossRef]
32. European Food Safety Authority Tolerable Upper Intake Levels for Vitamins and Minerals. Available online: http://www.efsa.europa.eu/sites/default/files/efsa_rep/blobserver_assets/ndatolerableuil.pdf (accessed on 3 May 2018).
33. Kader, A.A. Flavor quality of fruits and vegetables. *J. Sci. Food Agric.* **2008**, *88*, 1863–1868. [CrossRef]
34. Bornet, F.R.; Costagliola, D.; Rizkalla, S.W.; Blayo, A.; Fontvieille, A.M.; Haardt, M.J.; Letanoux, M.; Tchobroutsky, G.; Slama, G. Insulinemic and glycemic indexes of six starch-rich foods taken alone and in a mixed meal by type 2 diabetics. *Am. J. Clin. Nutr.* **1987**, *45*, 588–595. [CrossRef] [PubMed]
35. Beckles, D.M. Factors affecting the postharvest soluble solids and sugar content of tomato (*Solanum lycopersicum* L.) fruit. *Postharvest Biol. Technol.* **2012**, *63*, 129–140. [CrossRef]
36. George, B.; Kaur, C.; Khurdiya, D.S.; Kapoor, H.C. Antioxidants in tomato (*Lycopersium esculentum*) as a function of genotype. *Food Chem.* **2004**, *84*, 45–51. [CrossRef]
37. Lenucci, M.S.; Caccioppola, A.; Durante, M.; Serrone, L.; De Caroli, M.; Piro, G.; Dalessandro, G. Carotenoid content during tomato (*Solanum lycopersicum* L.) fruit ripening in traditional and high-pigment cultivars. *Ital. J. Food Sci.* **2009**, *21*, 461–472.
38. Ilahy, R.; Hdider, C.; Lenucci, M.S.; Tlili, I.; Dalessandro, G. Antioxidant activity and bioactive compound changes during fruit ripening of high-lycopene tomato cultivars. *J. Food Compos. Anal.* **2011**, *24*, 588–595. [CrossRef]
39. Hart, D.J.; Scott, K.J. Development and evaluation of an HPLC method for the analysis of carotenoids in foods, and the measurement of the carotenoid content of vegetables and fruits commonly consumed in the UK. *Food Chem.* **1995**, *54*, 101–111. [CrossRef]
40. Raiola, A.; Tenore, G.; Barone, A.; Frusciante, L.; Rigano, M. Vitamin E content and composition in tomato fruits: Beneficial roles and bio-fortification. *Int. J. Mol. Sci.* **2015**, *16*, 29250–29264. [CrossRef] [PubMed]
41. Seybold, C.; Frohlich, K.; Bitsch, R.; Otto, K.B.V. Changes in contents of carotenoids and vitamin E during tomato processing. *J. Agric. Food Chem.* **2004**, *5*, 7005–7010. [CrossRef] [PubMed]
42. Durante, M.; Montefusco, A.; Marrese, P.P.; Soccio, M.; Pastore, D.; Piro, G.; Mita, G.; Lenucci, M.S. Seeds of pomegranate, tomato and grapes: An underestimated source of natural bioactive molecules and antioxidants from agri-food by-products. *J. Food Compos. Anal.* **2017**, *63*, 65–72. [CrossRef]
43. Soto-Zamora, G.; Yahia, E.M.; Brecht, J.K.; Gardea, A. Effects of postharvest hot air treatments on the quality and antioxidant levels in tomato fruit. *LWT—Food Sci. Technol.* **2005**, *38*, 657–663. [CrossRef]
44. Harris, J.R. *Ascorbic Acid: Biochemistry and Biomedical Cell Biology*; Plenum Press: New York, NY, USA, 1996; ISBN 1461303257.

agriculture

MDPI

Article

Cultivation of Potted Sea Fennel, an Emerging Mediterranean Halophyte, Using a Renewable Seaweed-Based Material as a Peat Substitute

Francesco Fabiano Montesano [1], Concetta Eliana Gattullo [2], Angelo Parente [1,*], Roberto Terzano [2] and Massimiliano Renna [1,3]

[1]   Institute of Sciences of Food Production, CNR—National Research Council of Italy, Via G. Amendola, 122/O, 70126 Bari, Italy; francesco.montesano@ispa.cnr.it (F.F.M.); massimiliano.renna@uniba.it (M.R.)
[2]   Department of Soil, Plant and Food Sciences, University of Bari Aldo Moro, Via Amendola 165/A, 70126 Bari, Italy; concettaeliana.gattullo@uniba.it (C.E.G.); roberto.terzano@uniba.it (R.T.)
[3]   Department of Agricultural and Environmental Science, University of Bari Aldo Moro, Via Amendola, 165/A, 70126 Bari, Italy
*   Correspondence: angelo.parente@ispa.cnr.it; Tel.: +39-080-592-9309

Received: 18 May 2018; Accepted: 25 June 2018; Published: 27 June 2018

Abstract: Sea fennel (*Crithmum maritimum* L.), an emerging halophyte species, represents a nutritious and refined food product. In this study, the effect on yield and quality of potted sea fennel grown on three posidonia (*Podisonia oceanica* (L.) Delile)-based composts (a municipal organic solid waste compost, a sewage sludge compost and a green compost) and a peat-based substrate was analyzed. Composts were used both pure and mixed with peat at a dose of 50% on a volume basis. We hypothesized that the halophytic nature of this plant might overcome the limitations of high-salinity compost-based growing media. The growth parameters, color traits and trace metals content (Cd, Co, Cr, Cu, Fe, Mn, Ni, Pb and Zn) of the edible parts were compared. Independently of the substrates, the average total and edible yields were 51 and 30 g plant^{-1}, respectively, while the average waste portion was about 41%. The use of posidonia-based compost did not affect the color traits of sea fennel plants as compared with samples grown on the commercial peat-based substrate. In general, potted sea fennel grown on both posidonia-based composts and commercial peat-based substrate appeared a good source of essential micronutrients. Only a weak reduction of Fe and Mn concentrations was observed in plants grown on posidonia-based composts, especially when used at the highest dose. Independently of the growing medium, the content of potentially hazardous trace elements (Cd and Pb) in the edible parts of sea fennel was always below the maximum admissible limits fixed by the European legislation. Results indicate that posidonia-based composts can be used as a sustainable peat substitute for the formulation of soilless mixtures to grow potted sea fennel plants, even up to a complete peat replacement.

Keywords: *Crithmum maritimum* L.; domestication; food safety; heavy metal; *Posidonia oceanica* (L.) Delile; growing substrate

---

1. Introduction

Sea fennel (*Crithmum maritimum* L.) is a halophyte species belonging to the *Apiaceae* family, also known as crest marine, marine fennel and rock samphire because of its habit to grow as a wild plant on maritime rocks, breakwaters and sandy beaches. It is used as an ingredient of many dishes for its interesting sensory attributes like taste, odor and color [1,2]. The fresh leaves can be used to prepare salads, soups and sauces, or they are pickled in vinegar similarly to capers; this latter food preparation is listed as an item in the 'List of Traditional Agri-Food Products' of the Italian

Department for Agriculture [2], confirming the typical use of this plant in the Mediterranean tradition. However, also in British Isles, "Rock Samphire Hash" is a traditional recipe based on stems and leaves of *C. maritimum* L., mixed with pickled cucumbers and capers. Apart from the food use, sea fennel has been largely considered also for its nutritional and healthy properties [3]. In ancient times, this plant was used in traditional medicine for its stimulating, diuretic and vermifuge effects. In Italy, the sea fennel decoction was used against cystitis, prostatitis and colics, while the infusion was used in case of digestive diseases [4]. Despite the above mentioned notable characteristics, and the recent general interest for the recovery of traditional uses of plants, this species is currently considered an underutilized crop for commercial cultivation. Some Authors highlight benefits from cultivation and food use of a wide availability of genetic resources like wild edible plants, especially for their potential beneficial elements content [5–7]. To our best knowledge, the literature lacks information with regard to the sea fennel as a potential source of mineral elements in the daily diet. Moreover, only few information is available in the literature with regard to domestication and technical aspects of sea fennel growing techniques. Therefore, more information is needed in the view of a large-scale production of sea fennel, aimed to the optimal crop performance and, ultimately, to a sustainable exploitation of this emerging halophyte [2].

Like several other herbs (basil, parsley, chives), sea fennel may be suitable to be grown and potentially marketed as potted plant. In general, commercial production of potted plants implies the use of growing substrates other than real soil (soilless cultivation).

In southern Europe, the most used component for potting soilless substrates is peat, because of its good chemical and physical properties. However, peat use is becoming more and more problematic for a number of reasons, mainly related to its high costs (especially in Mediterranean countries, where peat is imported from extraction areas in Northern and Central Europe), and recently to the environmental concerns on peat extraction. Peat is a non-renewable material and its extraction can degrade wetland ecosystems, so that European policy strongly encourages the use of peat alternatives [8,9]. A recent EU Commission decision states that growing media (including soil improvers) containing peat material cannot receive the European Union "eco-label", thus encouraging the use of organic matter derived from the processing and/or reuse of wastes [10]. Among the organic materials proposed as alternatives to peat, different organic sources (e.g., municipal sewage sludge, organic fraction of urban solid wastes, food industry and wood processing residues and byproducts, agricultural residues) subjected to appropriate stabilization treatments, such as composting, are arising interest as components of soilless growing substrates. A number of experimental evidences demonstrate that, if properly formulated, compost-based growing substrates may offer interesting possibilities for the reduction of peat use, although it is generally recommended not to exceed a 50% rate of compost in the mixture, in order to avoid a reduction of the crop performance due to reduced plant growth or possible phytotoxicity problems [11]. On the other hand, several investigations focused on the possible risks for human health of potentially toxic elements (PTEs) accumulation in vegetables, associated with the use of compost-based growing substrates [12,13].

Posidonia (*Posidonia oceanica* (L.) Delile) is the most important endemic marine phanerogam in the Mediterranean Sea [14]. As a part of its natural life cycle, this plant loses its senescent parts, which accumulate in large quantities along the coasts, often originating problems related to the management of this organic material. Composting may be a viable alternative to landfill disposal for this material [15]. The posidonia-based compost has been proposed as a promising soilless growing media component [16]. Posidonia compost-based substrates have been successfully tested to produce tomato [17–19], lettuce [12,20], potted basil [13], and to grow transplant seedlings of lettuce [21], melon and tomato [22]. However, no information is available about the use of posidonia-based composts as growing substrate for potted sea fennel cultivation.

Moving from these considerations, the objective of the present study was to ascertain the potential use of posidonia compost-based substrates for potted sea fennel production. We focused on: (i) the evaluation of the yield and quality of sea fennel grown on different posidonia-based compost soilless substrates, using a commercial peat-based substrate as a reference; and (ii) the accumulation of PTEs in the edible parts of the plant.

2. Materials and Methods

2.1. Starting Materials for Growing Media Mixtures

A commercial peat-based substrate (**P**) and three different compost-based materials (**Gr C**, **SS C** and **MOW C**) containing posidonia residues in their formulation, were used as components for the preparation of the growing media tested in the present study (see Section 2.2 for a detailed description of the mixtures). 'P' was a commercial horticultural substrate based on a mix of different peats enriched with 1 kg m^{-3} PG-MIX 14-16-18 fertilizer (Brill Type 3 Special). '**MOW C**', hereafter referred to as municipal organic solid waste compost was obtained by mixing posidonia residues ($\approx$16% on a fresh weight basis), organic fraction of municipal solid wastes ($\approx$29%), urban green wastes ($\approx$27%), and agro-industrial sludge from tomato, grape and olive processing (28%). '**SS C**', hereafter referred to as sewage sludge compost, was obtained by mixing posidonia residues ($\approx$16%), urban sewage sludge ($\approx$29%), urban green wastes (pruning and yard trimming residues) ($\approx$27%), and agro-industrial sludge (28%). '**Gr C**', hereafter referred to as green compost, was obtained by mixing posidonia residues ($\approx$20% of the composting pile, on a fresh weight basis), green residues resulting from a greenhouse tomato cultivation ($\approx$40%), and pruning residues of an olive grove ($\approx$40%). Composts were produced in a local industrial plant according to the Italian regulations on compost production. The production process was carried out in piles over three months. Piles were mechanically turned in accordance with temperature evolution. The principal chemical features of the materials are reported in Table 1.

2.2. Plant Material and Experimental Conditions

The trial was conducted in a polymethacrylate greenhouse located in Mola di Bari (Bari, Italy) at 'La Noria' experimental farm of the CNR-ISPA (17°04' E, 41°03' N, 24 m a.s.l.). Average daily mean, minimum and maximum air temperatures inside the greenhouse over the experiment were 19.2, 11.9 and 29.7 °C, respectively. Average daily mean, minimum and maximum air relative humidity values were 65, 36 and 87%, respectively.

Seven mixtures were prepared starting from the above described materials, and tested as growing media. Each of the posidonia-based composts (Gr C, SS C and MOW C) was used both pure, at a rate of 100%, and mixed with P at rate of 50% on a volume basis. A control treatment, consisting only of pure peat-based commercial substrate (P), was also included. The resulting seven treatments under comparison in the study were the following: (i) P, (ii) MOW C, (iii) MOW C + P, (iv) SS C, (v) SS C + P, (vi) Gr C, (vii) Gr C + P. The electrical conductivity (EC) and pH of the seven substrates, both measured on the 1:10 (*w/v*) aqueous extracts, were: P (EC = 0.3 dS m^{-1}; pH = 6.8); MOW C (EC = 2.9 dS m^{-1}; pH = 7.8), MOW C + P (EC = 1.5 dS m^{-1}; pH = 7.5), SS C (EC = 2.3 dS m^{-1}; pH = 7.6), SS C + P (EC = 1.6 dS m^{-1}; pH = 7.5), Gr C (EC = 1.7 dS m^{-1}; pH = 8.3), Gr C + P (EC = 1.1 dS m^{-1}; pH = 7.4).

Seeds of *C. maritimum* L. were harvested from wild plants along the shoreline in Mola di Bari. Ascorbic acid at 40 mM was used as pre-treatment to improve sea fennel seed germination, according to the procedure described by Meot-Duros and Magné [23]. The seedlings were produced in polystyrene plug trays (160 cells per tray with diameter of 2.5 cm and volume of 21 mL) filled with peat. The seeds, three per cell, were covered with vermiculite. Seven days after emergence, the number of plants was reduced to one seedling per cell. After growing in conventional plug trays for 30 days, seedlings were transferred into 10 cm diameter plastic containers (0.5 L) filled with each of the seven growing media. The pots were placed on benches and grown in a closed cycle ebb and flow hydroponic system. The experimental treatments were organized in a fully randomized design with three replications for each treatment, each replication constituted by nine pots, for a total of 189 pots in the experiment (27 per treatment).

Table 1. Main chemical features of the materials used for the preparation of the growing media mixtures.

	C/N	P	Ca	K	Mg	Na	Fe	Cu	Mn	Ni	Zn	B	Cd	Cr	Cr(VI)	Pb	Hg
	Ratio	g kg⁻¹ DW						mg kg⁻¹ DW									
P [1]	-	1.41	22.0	1.27	1.59	0.28	1.13	6.7	72	5.0	9	14	0.07	1.4	-	2.6	-
MOW C	11.6	4.23	17.6	7.58	1.92	1.21	2.58	52.9	156	4.4	120	24	<0.50	6.8	<0.50	7.3	<0.50
SS C	12.5	5.08	71.0	5.14	3.18	1.19	5.97	89.7	235	11.5	170	94	<0.50	18.9	<0.50	18.8	<0.50
Gr C	11.8	0.75	16.6	1.39	1.91	1.49	1.31	8.9	29	3.9	43	158	<0.50	1.4	<0.50	2.2	<0.50
Maximum Limits According to Regulations																	
Italy [2]	25	-	-	-	-	-	-	230		100	500		1.5	-	0.5	140	1.5
European Union [3]	-	-	-	-	-	-	-	100		50	300		1.0	100	-	100	1.0

P: Peat-based commercial substrate; MOW C: Municipal organic solid waste compost; SS C: Sewage sludge compost; Gr C: Green compost. [1] Adapted from Minimni et al. [13]. [2] Italian directive on fertilizers (D.L. 75/2010, Annex 2). [3] Ecological criteria for EU eco-label [10].

Plants were grown for 60 days with nutrient solution (NS) containing N (140 mg L^{-1}, NO_3-N:NH_4-N 80:20), P (50 mg L^{-1}), K (200 mg L^{-1}), Mg (40 mg L^{-1}) and Ca (100 mg L^{-1}), as macronutrients, whereas micronutrients were supplied according to Johnson et al. [24]. The NS pH was adjusted to 5.5–6.0 using 1 M H_2SO_4. The well water used to prepare the NS had the following characteristics: pH 7.1, EC 1.1 dS m^{-1}, 1.0 M Ca^{2+}, 0.4 M Mg^{2+}, 6.0 M Na^+ and 5.7 M Cl^-. The chemical composition of well water was taken into account for the preparation of the NS. Benches were flooded with the NS three times per day. The NS level in the benches was raised up to about 3 cm from the bottom of the pots, allowing the substrates to absorb the NS for about 10 min per irrigation event before the NS was discharged and collected for subsequent irrigations.

2.3. Plant Growth and Color Parameters Measurements

At harvest, yield and leaf area were measured on six plants per replicate. Yield was expressed in terms of both total and edible (after the removal of waste portion) fresh weight of shoots per plant. The waste portion consisted of older leaves and stems that are generally removed during the normal food use, and was expressed as percentage of the total shoots biomass. Leaf area was measured with a leaf area meter (LI-3100, LI-COR, Lincoln, NE, USA). For each sample, a portion of the plant material was dried in a forced draft oven at 105 °C until reaching a constant weight for the determination of the dry matter content, which was expressed as g 100 g^{-1} fresh weight (FW).

Color measurements were performed using a colorimeter (CR-400, Konica Minolta, Osaka, Japan) equipped with illuminant D65, operating in reflectance mode and with the CIE L (lightness) a^* (redness) b^* (yellowness) color scale, according to the procedure described by Montesano et al. [25]. The colorimeter was preliminary calibrated with a standard reference material characterized by L, a^* and b^* values of 97.55, 0.52 and 1.45, respectively. Hue angle ($h° = arctg$ b^*/a^*) and saturation ($C = [a^{*2} + b^{*2}]^{1/2}$) were then calculated from primary L, a^* and b^* readings.

2.4. Elemental Analysis

A representative aliquot of each plant sample was oven-dried at 65 °C and finely ground using a cutting-grinding mill (IKA MF 10B, Labortechnik, Staufen, Germany), then a subsample of 0.3 g was pre-digested overnight with 7 mL HNO_3 (69%) and 1 mL H_2O_2 (30%) (TraceSELECT®, trace analysis reagents, Sigma Aldrich, St. Louis, MO, USA) in a PTFE-TFM liner. Sample digestion was performed using a microwave oven (Multiwave 3000, Anton Paar, Graz, Austria), according to the procedure described by Gattullo et al. [12]. Digested samples were diluted with deionised water up to 25 mL, then filtered through Whatman® 42 filter paper and stored at 4 °C until analysis. All glassware was cleaned using 5% HNO_3 solution, and then abundantly rinsed with deionised water. Total concentrations of Fe, Mn, Cu, Zn, Cd, Cr, Co, Ni and Pb were determined by inductively coupled plasma atomic emission spectrometry (ICP-AES; Thermo iCAP 6000 series, Thermo Fisher Scientific Inc., Waltham, MA, USA), following the method described by Gattullo et al. [12]. A two-point calibration was adopted using the blank (HNO_3/H_2O_2 7:1, v/v) as zero point, and a multi-element calibration standard (Certipur® ICP Multi-element standard solution IV; Merck Millipore, Burlington, MA, USA) at the concentration of 2 mg L^{-1} (prepared in the blank acidic solution). Instrument detection limits were calculated for each element as three times the standard deviation of ten replicates of the blank.

2.5. Statistical Analysis

Data were subjected to one-way analysis of variance (ANOVA). Treatment means were separated by LSD test when there was a significant effect at the $p < 0.05$ level. The statistical software STATISTICA 10.0 (StatSoft, Tulsa, OK, USA) was used for the analysis.

3. Results and Discussion

3.1. Yield, Leaf Area, Dry Matter and Color Parameters

The average total and edible yields were 51 and 30 g plant^{-1}, respectively, while the average waste portion was about 41%, without any significant difference between the growing substrates under comparison (Table 2). However, although not statistically significant, a certain decrease in yield parameters was observed in MOW C treatment. No significant differences were found between the substrates also for the leaf area and dry matter content, with an average value of 296 cm^2 plant^{-1} and 10.7 g 100 g^{-1} FW, respectively (Table 2). No visual symptoms of nutritional deficiencies or toxicities were observed on plants during the overall growing cycle. These results are in agreement with Parente et al. [16], and reveal that posidonia-based composts may be suitable materials for the formulation of soilless mixtures to grow containerized sea fennel plants, without any negative effect on growth parameters. A similar growth response was obtained using composts both at 100% and 50% rate in mixture with peat (Table 2), in contrast with the general recommendation to not exceed a 50% rate [21,22]. One of the main problems related to the use of composts is generally their high salt content, as outlined also for posidonia-based composts [18]. Indeed, the composts used in this study presented a higher salt content in comparison with the traditional peat-based commercial substrate. Despite of this, the three composts did not reduce sea fennel growth even when used at the higher dose (Table 2), possibly because of the halophytic behavior of this plant species [26]. This result further supports the use of posidonia-based composts as a potential peat substitute for the containerized sea fennel production, even up to a complete peat replacement. By considering the general interest on alternative soilless materials, and the particular attention devoted to organic materials derived from waste streams [27], the findings of this study may represent an important scientific background to develop a sustainable sea fennel production. In particular, the use of composted posidonia residues as a renewable, low-cost and locally available material, is especially feasible in areas where sea fennel cultivation may represent an opportunity for growers, as the coastal areas of Mediterranean countries [16]. Moreover, the sustainability of the cultivation process is further increased by using the ebb and flow technique, a typical soilless closed-cycle subirrigation system, which prevent pollution, while reducing water and fertilizer use [28]. Among the different soilless systems based on the closed-cycle management of the NS, subirrigation for containerized crops has been recognized as a practical and cost-effective method to achieve efficient water use in the Mediterranean greenhouse industry, due to the relative simplicity of the subirrigation techniques [29].

Table 2. Total shoot fresh weight (FW), edible yield, waste portion, leaf area and dry matter of sea fennel grown on seven substrates composed of peat and three different composts containing posidonia residues. A peat-based commercial substrate was used as control. The compost materials were tested alone or in mixture with peat (at a rate of 50% on a volume basis).

	Total FW	Edible Yield	Waste Portion	Leaf Area	Dry Matter
	g plant^{-1}			cm^2 plant^{-1}	g 100 g^{-1} FW
P	58.6 ± 9.8	35.1 ± 6.0	40.0 ± 3.1	343 ± 59	11.4 ± 1.1
MOW C	34.5 ± 13.0	21.0 ± 7.4	38.4 ± 3.1	216 ± 8	10.5 ± 0.3
MOW C + P	47.5 ± 5.6	27.2 ± 4.9	42.9 ± 3.5	278 ± 32	10.7 ± 0.8
SS C	45.0 ± 3.6	26.3 ± 2.5	41.6 ± 3.5	258 ± 36	10.7 ± 0.9
SS C + P	54.3 ± 9.8	31.5 ± 6.1	40.6 ± 6.9	302 ± 70	10.3 ± 0.6
Gr C	54.5 ± 15.2	30.4 ± 6.7	43.5 ± 4.2	307 ± 70	10.2 ± 0.8
Gr C + P	61.9 ± 7.5	36.4 ± 5.3	41.3 ± 2.7	369 ± 37	11.0 ± 1.4
Significance	NS	NS	NS	NS	NS

P: Peat-based commercial substrate; MOW C: Municipal organic solid waste compost; SS C: Sewage sludge compost; Gr C: Green compost. NS = not significant. Values are the mean of three replications (six subsamples per replication), ± standard deviation.

On the basis of the shoots dry matter, the total average sea fennel biomass production was about 5.4 g dry weight (DW) plant^{-1} (Table 2). These results are in agreement with Hamed et al. (2004) who reported a potted sea fennel biomass production between 1 and 13 g DW plant^{-1} and a leaf area between about 100 and more than 600 cm^2 plant^{-1}, at NaCl concentration in the NS ranging from 0 to 300 mM. A lower sea fennel biomass production, however, was found in other studies, with values ranging from 2.1 to 3.0 g DW plant^{-1} [26,29].

As regard the color analysis of the sea fennel leaves, no significant differences were found between the different treatments (Table 3) with an average *L* (lightness), *a** (redness), *b** (yellowness), h° (Hue angle) and C (saturation) values of 41.5, −11.5, 16.2, 125.3 and 19.9, respectively. This indicates that the color of sea fennel grown on different posidonia-based composts is similar to that obtained by using commercial peat-based substrates. These results are in agreement with Renna et al. [2], who found similar color parameters on wild sea fennel collected along sea shoreline, which is the natural habitat of this species. Color is among the first quality parameters catching the attention of consumers, and exerts a strong influence on consumers' choice and opinion about the food quality [30]. Moreover, color is one of the most important quality traits of the sea fennel [1,2].

Table 3. Color parameters [*L* (lightness), *a** (redness), *b** (yellowness), h° (Hue angle) and C (saturation)] of sea fennel grown on seven substrates composed of peat and three different composts containing posidonia residues. A peat-based commercial substrate was used as control. The compost materials were tested alone or in mixture with peat (at a rate of 50% on a volume basis).

	L	*a**	*b**	h°	C
P	41.8 ± 1.2	−11.4 ± 0.5	16.2 ± 0.8	125.3 ± 0.2	19.8 ± 0.9
MOW C	41.7 ± 0.8	−11.6 ± 0.1	16.3 ± 0.4	125.2 ± 0.4	19.9 ± 0.3
MOW C + P	41.4 ± 0.6	−11.6 ± 0.3	16.5 ± 0.4	125.1 ± 0.2	20.1 ± 0.6
SS C	41.3 ± 0.5	−11.5 ± 0.4	16.3 ± 0.6	125.2 ± 0.4	19.9 ± 0.7
SS C + P	41.5 ± 0.6	−11.2 ± 0.5	15.9 ± 0.2	125.2 ± 0.8	19.5 ± 0.5
Gr C	41.1 ± 0.3	−11.7 ± 0.4	16.4 ± 0.7	125.5 ± 0.5	20.2 ± 0.8
Gr C + P	41.7 ± 1.2	−11.5 ± 0.2	16.2 ± 0.6	125.4 ± 0.6	19.8 ± 0.6
Significance	NS	NS	NS	NS	NS

P: Peat−based commercial substrate; MOW C: Municipal organic solid waste compost; SS C: Sewage sludge compost; Gr C: Green compost. NS = not significant. Values are the mean of three replications (six subsamples per replication), ± standard deviation.

3.2. Trace Metals Content

Table 4 shows the concentration of some trace metal elements in potted sea fennel grown on different substrates. Cadmium, Co, Ni and Pb concentrations in plant tissues were below the instrument detection limits of 0.002 (Cd), 0.015 (Co), 0.193 (Ni) and 0.063 (Pb) mg kg^{-1} FW (data not shown). Among these elements, Cd and Pb are considered highly toxic to humans, therefore the European Commission fixed the maximum admissible concentration in vegetables at 0.1 and 0.05 mg kg^{-1} FW for Pb and Cd, respectively [10]. Cadmium intake through food can disturb the activity of several enzymes and proteins due to Cd inactivation of sulfhydryl groups, and cause kidney and liver dysfunction, as well as osteoporosis and other pathologies [31,32]. Lead is a neurotoxic element, and its excess in human body may also induce cardiovascular problems, nephrotoxicity, as well as carcinogenicity and genotoxicity [32,33]. In this study, the levels of Cd and Pb in sea fennel plants grown on posidonia−based compost substrates were below the limits imposed by the legislation and therefore safe for human consumption. In fact, the three composts used for sea fennel growth complied with the maximum limits fixed by the European and Italian legislation for PTEs in fertilizers (Table 1) and therefore, in this respect, they were safe for plant growth.

Table 4. Trace metals tissue concentration of sea fennel grown on seven substrates composed of peat and three different composts containing posidonia residues. A peat−based commercial substrate was used as control. The compost materials were tested alone or in mixture with peat (at a rate of 50% on a volume basis).

	Cr	Cu	Fe	Mn	Zn
			mg kg^{-1} DW		
P	0.22 ± 0.05	2.33 ± 0.43	57.05 ± 11.29 a	26.07 ± 3.38 a	13.40 ± 3.24
MOW C	0.19 ± 0.03	2.77 ± 0.17	40.21 ± 3.31 b	18.11 ± 1.44 cd	17.62 ± 5.69
MOW C + P	0.17 ± 0.05	2.45 ± 0.87	49.50 ± 8.05 ab	21.49 ± 1.59 bc	12.69 ± 5.14
SS C	0.18 ± 0.05	2.92 ± 0.66	42.22 ± 2.75 b	18.73 ± 0.78 cd	15.58 ± 2.52
SS C + P	0.18 ± 0.02	2.72 ± 0.39	48.52 ± 1.11 ab	25.43 ± 0.76 ab	15.80 ± 2.06
Gr C	0.22 ± 0.08	2.52 ± 0.37	42.11 ± 2.92 b	23.83 ± 5.26 ab	16.59 ± 3.97
Gr C + P	0.18 ± 0.03	2.17 ± 0.30	45.58 ± 3.12 b	15.48 ± 1.21 d	15.64 ± 4.21
Significance	NS	NS	*	**	NS

P: Peat−based commercial substrate; MOW C: Municipal organic solid waste compost; SS C: Sewage sludge compost; Gr C: Green compost. *, ** significant at $p \leq 0.01$, and $p \leq 0.001$, respectively; NS = not significant. Values are the mean of three replications (six subsamples per replication), ±standard deviation. For each column different letters indicate statistically significant differences at $p = 0.05$.

The levels of total Cr in sea fennel plants were not influenced by the growing substrate (Table 4). Chromium is not an essential nutrient for plants, whereas it is an important micronutrient for humans, as it regulates the metabolism of glucose and lipids. Despite the lack of law restrictions for Cr content in vegetables, high concentrations of this element are not desired due to the carcinogenic effect of its hexavalent form [32]. In this study, the values of total Cr concentrations were similar to those reported for other vegetables grown on compost−based substrates, such as lettuce [12] and basil [13], or cultivated on soil [34]. Moreover, as reported in Table 1, Cr(VI) concentrations in the three composts were below the legislation limit, thus it can be expected that Cr was accumulated in plants as Cr(III), which is the non−toxic form.

With regard to micronutrients, the growing substrate influenced only the accumulation of Fe and Mn, whereas Cu and Zn were not affected (Table 4). Iron concentration in sea fennel grown on P was significantly higher with respect to plants grown on MOW C, SS C, Gr C and Gr C + P, but not significantly different from plants grown on MOW C + P and SS C + P (Table 4). As regard Mn, its concentration in sea fennel grown on P was significantly higher respect to all other treatments with the exception of SS C + P and Gr C treatments (Table 4). Similarly to our findings, a general higher concentration of Fe and Mn was observed in basil [13] and lettuce [12] grown on peat compared to plants grown on substrates containing posidonia−based composts. Despite the total Fe and Mn content in MOW C and SS C was much higher than in P (Table 1), the higher pH and salinity of the two compost−based substrates possibly reduced the uptake and accumulation of these nutrients by plants, compared to the control (P). However, when MOW C and SS C were used at 50% rate, the overall substrate salinity declined producing beneficial effects on element accumulation in plant shoots (with the exception of Mn for the treatment MOW C + P). This beneficial effect was not observed in plants grown on Gr C + P, possibly because the total content of Fe and Mn in Gr C was not high enough to balance the negative effect induced by pH and salinity.

3.3. Potential Human Intake of Essential Micronutrients

In recent years, there is a growing interest for the evaluation of mineral element content in vegetables due to their nutritional properties and health effects [35]. Some of these elements occur in human body at trace concentrations (milligrams or micrograms per kilogram of tissue) and are considered "essential" since a deficient intake may consistently result in an impairment of biological functions from optimal to suboptimal [36]. To the best of our knowledge, the literature lacks information with regard to sea fennel as a potential source of mineral elements in the daily diet.

Therefore, the intake of some essential trace elements for a serving size of potted sea fennel has been evaluated, like previously done for other domesticated and wild edible plants used to prepare dishes [5].

Regarding Cr, there is no formal Recommended Dietary Allowance (RDA). Nevertheless, the US Food and Nutrition Board [37] derived Adequate Intakes (AI) of 35 and 25 μg day^{-1} for 19–50 year old men and women, respectively. For a serving size of 100 g FW, independently of the used growing media, a portion of sea fennel could supply, on average, 2.1 μg Cr that is about the 6% and 8.4% of Cr daily AI for men and women, respectively. Anderson et al. [38] reported for other vegetables a Cr content between 0.4 and 1.6 μg 100 g^{-1} FW. Therefore, potted sea fennel may be considered a good source of this essential element.

The content of Cu in potted sea fennel was about 28 μg 100 g^{-1} FW, independently of the used growing media. Copper is an essential trace element for living organisms, including humans, being a vital component of several enzymes and proteins [39]. The results of the present study show a very low Cu content in potted sea fennel, considering its recommended dietary intake of 900 μg day^{-1} [40].

Regarding Fe content, 100 g FW of sea fennel grown on P contains 0.64 mg of this element, supplying about 7% of the recommended dietary intake [37], while in all other cases a serving size contains, on average, 0.47 mg Fe 100 g^{-1} FW (about 5% of the recommended dietary intake), independently of the used growing media. Iron is an important trace element for human body, since it acts as oxygen carrier in hemoglobin in blood and myoglobin in muscle. Iron deficiency is the only nutrient deficiency which is also significantly prevalent in industrialized countries [41]. It is important to highlight that the main Fe sources in the human diet are meat and other animal products, which provide more bioavailable Fe than vegetables. However, some leafy vegetables such as spinach can moderately contribute to human Fe intake. The results of the present study show that the Fe content in sea fennel is lower respect to spinach but similar or higher in comparison with other vegetables, such as chicory and iceberg lettuce, respectively [42].

As for Mn, potted sea fennel contains between 0.2 and 0.3 mg Mn 100 g^{-1} FW depending on the growing media. Manganese plays an important role in several physiological functions of the human body, and the Food and Nutrition Board [37] recommended an adequate intake of about 2.0 mg day^{-1}. Therefore, a serving size of potted sea fennel could supply between 10 and 15% of the Mn daily intake.

Zinc content in potted sea fennel was about 0.16 mg 100 g^{-1} FW, independently of the used growing media. Zinc is an essential trace element for humans since it takes part in many enzymatic reactions, and plays an important role in growth and development, immune response, neurological function, and reproduction [43]. Considering that the European Population Reference Intake (PRI) for Zn is 9.5 and 7.0 mg day^{-1} for adult males and females, respectively [40], the results of the present study show a low Zn content in potted sea fennel, independently of the used growing media. On the other hand, it is important to highlight that the Zn content in vegetables is usually below 1 mg 100 g^{-1} FW and in many types of vegetables, such as cabbage, cardoon, celery and lettuce, its content is similar to what found in potted sea fennel [42].

4. Conclusions

In this study the potential use of posidonia compost–based substrates for potted sea fennel production was evaluated. Our results show the possibility to use these substrates without any negative effect on the sea fennel growth in comparison with a commercial peat substrate. The halophytic nature of this plant may have played a role in overcoming the limitations posed by the high–salinity of compost–based growing substrates. Also for the color parameters, which are among the most important quality traits of the sea fennel, no difference was observed between plants grown on posidonia–based composts and a commercial peat–based substrate. The accumulation of elements potentially toxic for human health in the edible parts of the plant was far below the limits imposed by legislation and therefore safe for human consumption. At the same time, potted sea fennel grown on both posidonia–based composts and commercial peat–based substrate may be considered a good source of some essential micronutrients for humans. Therefore, posidonia−based composts can be used as sustainable peat substitutes for the formulation of soilless mixtures to grow potted sea fennel plants, even up to a complete peat replacement. The use of composted posidonia residues as a renewable and low−cost material could be feasible especially in coastal areas of Mediterranean countries. Finally, these results can serve as a basis for the cultivation of other emerging halophyte species on posidonia−based compost substrates.

Author Contributions: Conceptualization, F.F.M., A.P. and M.R.; Investigation, F.F.M., C.E.G, A.P., R.T. and M.R.; Formal Analysis, A.P. and M.R.; Writing−Original Draft Preparation, F.F.M. and M.R.; Writing−Review & Editing, F.F.M., C.E.G., A.P., R.T. and M.R.; Supervision, M.R.; Funding Acquisition, M.R.

Funding: This research was funded by Regione Puglia Administration under "Intervento cofinanziato dal Fondo di Sviluppo e Coesione 2007−2013—APQ Ricerca Regione Puglia—Programma regionale a sostegno della specializzazione intelligente e della sostenibilità sociale ed ambientale FutureInResearch"—project 'Innovazioni di prodotto e di processo per la valorizzazione della Biodiversità Orticola pugliese (InnoBiOrt)'.

Acknowledgments: The authors thank Nicola Gentile for technical assistance.

Conflicts of Interest: The authors declare no conflict of interest.

References

1. Renna, M.; Gonnella, M. The use of the sea fennel as a new spice-colorant in culinary preparations. *Int. J. Gastron. Food Sci.* **2012**, *1*, 111–115. [CrossRef]

2. Renna, M.; Gonnella, M.; Caretto, S.; Mita, G.; Serio, F. Sea fennel (*Crithmum maritimum* L.): From underutilized crop to new dried product for food use. *Genet. Resour. Crop Evol.* **2017**, *64*, 205–216. [CrossRef]

3. Pistrick, K. Current taxonomical overview of cultivated plants in the families Umbelliferae and Labiatae. *Genet. Resour. Crop Evol.* **2002**, *49*, 211–221. [CrossRef]

4. Atia, A.; Barhoumi, Z.; Mokded, R.; Abdelly, C.; Smaoui, A. Environmental eco-physiology and economical potential of the halophyte *Crithmum maritimum* L. (Apiaceae). *J. Med. Plants Res.* **2011**, *5*, 3564–3571.

5. Renna, M.; Cocozza, C.; Gonnella, M.; Abdelrahman, H.; Santamaria, P. Elemental characterization of wild edible plants from countryside and urban areas. *Food Chem.* **2015**, *177*, 29–36. [CrossRef] [PubMed]

6. Pignone, D.; Hammer, K. Parasitic angiosperms as cultivated plants? *Genet. Resour. Crop Evol.* **2016**, *63*, 1273–1284. [CrossRef]

7. Petropoulos, S.A.; Karkanis, A.; Martins, N.; Ferreira, I.C.F.R. Edible halophytes of the Mediterranean basin: Potential candidates for novel food products. *Trends Food Sci. Technol.* **2018**, *74*, 69–84. [CrossRef]

8. Bustamante, M.A.; Paredes, C.; Moral, R.; Agulló, E.; Pérez-Murcia, M.D.; Abad, M. Composts from distillery wastes as peat substitutes for transplant production. *Resour. Conserv. Recycl.* **2008**, *52*, 792–799. [CrossRef]

9. Grigatti, M.; Giorgioni, M.; Ciavatta, C. Compost-Based Growing Media: Influence on Growth and Nutrient Use of Bedding Plants. *Bioresour. Technol.* **2007**, *98*, 3526–3534. [CrossRef] [PubMed]

10. EU Commission Commission Decision of 3 November 2006 Establishing Revised Ecological Criteria and the Related Assessment and Verification Requirements for the Award of the Community Eco-Label to Growing Media (2006/799/EC). Available online: http://eur-lex.europa.eu/legal-content/EN/TXT/HTML/?uri=CELEX:32006D0799&from=EN (accessed on 4 May 2018).

11. Raviv, M. The future of composts as ingredients of growing media. *Acta Hortic.* **2011**, 19–32. [CrossRef]

12. Gattullo, C.E.; Mininni, C.; Parente, A.; Montesano, F.F.; Allegretta, I.; Terzano, R. Effects of municipal solid waste- and sewage sludge-compost-based growing media on the yield and heavy metal content of four lettuce cultivars. *Environ. Sci. Pollut. Res.* **2017**, *24*, 25406–25415. [CrossRef] [PubMed]

13. Mininni, C.; Grassi, F.; Traversa, A.; Cocozza, C.; Parente, A.; Miano, T.; Santamaria, P. *Posidonia oceanica* (L.) based compost as substrate for potted basil production. *J. Sci. Food Agric.* **2015**, *95*, 2041–2046. [CrossRef] [PubMed]

14. Boudouresque, C.F.; Bernard, G.; Bonhomme, P.; Groupement d'intéret scientifique Posidonie (Marseille). *Préservation et Conservation des Herbiers à Posidonia Oceanica*; Ramoge Pub: Monte Carlo, Monaco, 2006; ISBN 2905540303.

15. Parente, A.; Montesano, F.F.; Lomoro, A.; Guido, M. Improvement of Beached Posidonia Residues Performance to Composting. *Environ. Eng. Manag. J.* **2013**, *12*, 81–84.
16. Parente, A.; Serio, F.; Montesano, F.F.; Mininni, C.; Santamaria, P. The compost of Posidonia residues: A short review on a new component for soilless growing media. *Acta Hortic.* **2014**, 291–298. [CrossRef]
17. Castaldi, P.; Melis, P. Growth and Yield Characteristics and Heavy Metal Content on Tomatoes Grown in Different Growing Media. *Commun. Soil Sci. Plant Anal.* **2004**, *35*, 85–98. [CrossRef]
18. Montesano, F.F.; Parente, A.; Grassi, F.; Santamaria, P. Posidonia-based compost as a growing medium for the soilless cultivatipn of tomato. *Acta Hortic.* **2014**, 277–282. [CrossRef]
19. Verlodt, H.; Ben Abdallah, A.; Harbaoui, Y. Possibility of reutilization of a composted substrate of *Posidonia oceanica* (L.) Del. in a tomanto growth bag. *Acta Hortic.* **1984**, 439–448. [CrossRef]
20. Gizas, G.; Tsirogiannis, I.; Bakea, M.; Mantzos, N.; Savvas, D. Impact of hydraulic characteristics of raw or composted posidonia residues, coir, and their mixtures with pumice on root aeration, water availability, and yield in a lettuce crop. *HortScience* **2012**, *47*, 896–901.
21. Mininni, C.; Santamaria, P.; Abdelrahman, H.M.; Cocozza, C.; Miano, T. Posidonia-based compost as a peat substitute for lettuce transplant production. *HortScience* **2012**, *47*, 1438–1444.
22. Mininni, C.; Bustamante, M.A.; Medina, E.; Montesano, F.; Paredes, C.; Pérez-Espinosa, A.; Moral, R.; Santamaria, P. Evaluation of posidonia seaweed-based compost as a substrate for melon and tomato seedling production. *J. Hortic. Sci. Biotechnol.* **2013**, *88*, 345–351. [CrossRef]
23. Meot-Duros, L.; Magné, C. Effect of salinity and chemical factors on seed germination in the halophyte *Crithmum maritimum* L. *Plant Soil* **2008**, *313*, 83–87. [CrossRef]
24. Johnson, C.M.; Stout, P.R.; Broyer, T.C.; Carlton, A.B. Comparative chlorine requirements of different plant species. *Plant Soil* **1957**, *8*, 337–353. [CrossRef]
25. Montesano, F.F.; D'Imperio, M.; Parente, A.; Cardinali, A.; Renna, M.; Serio, F. Green bean biofortification for Si through soilless cultivation: Plant response and Si bioaccessibility in pods. *Sci. Rep.* **2016**, *6*, 1–9. [CrossRef] [PubMed]
26. Amor, N.B.; Hamed, K.B.; Debez, A.; Grignon, C.; Abdelly, C. Physiological and antioxidant responses of the perennial halophyte *Crithmum maritimum* to salinity. *Plant Sci.* **2005**, *168*, 889–899. [CrossRef]
27. Barrett, G.E.; Alexander, P.D.; Robinson, J.S.; Bragg, N.C. Achieving environmentally sustainable growing media for soilless plant cultivation systems—A review. *Sci. Hortic.* **2016**, *212*, 220–234. [CrossRef]
28. Ferrarezi, R.S.; Van Iersel, M.W.; Testezlaf, R. Monitoring and Controlling Ebb-and- flow Subirrigation with Soil Moisture Sensors. *HortScience* **2015**, *50*, 447–453.
29. Bouchaaba, Z.; Santamaria, P.; Choukr-Allah, R.; Lamaddalena, N.; Montesano, F.F. Open-cycle drip vs. closed-cycle subirrigation: Effects on growth and yield of greenhouse soilless green bean. *Sci. Hortic.* **2015**, *182*, 77–85. [CrossRef]
30. D'Imperio, M.; Renna, M.; Cardinali, A.; Buttaro, D.; Serio, F.; Santamaria, P. Calcium biofortification and bioaccessibility in soilless "baby leaf" vegetable production. *Food Chem.* **2016**, *213*, 149–156. [CrossRef] [PubMed]
31. Godt, J.; Scheidig, F.; Grosse-Siestrup, C.; Esche, V.; Brandenburg, P.; Reich, A.; Groneberg, D.A. The toxicity of cadmium and resulting hazards for human health. *J. Occup. Med. Toxicol.* **2006**, *1*, 1–6. [CrossRef] [PubMed]
32. Kabata-Pendias, A.; Mukherjee, A.B. *Trace Elements from Soil to Human*; Springer-Verlag: Berlin, Germany, 2007.
33. EU Commission Lead in Food. Available online: https://ec.europa.eu/food/safety/chemical_safety/contaminants/catalogue/lead_en (accessed on 5 May 2018).
34. Lendinez, E.; Lorenzo, M.L.; Cabrera, C.; López, M.C. Chromium in basic foods of the Spanish diet: Seafood, cereals, vegetables, olive oils and dairy products. *Sci. Total Environ.* **2001**, *278*, 183–189. [CrossRef]
35. Renna, M. Wild edible plants as a source of mineral elements in the daily diet. *Prog. Nutr.* **2017**, *19*, 219–222. [CrossRef]
36. Mertz, W. Human requirements: Basic and optimal. *Ann. N. Y. Acad. Sci.* **1972**, *199*, 191–201. [CrossRef] [PubMed]
37. Food and Nutritional Board. *Dietary Reference Intakes for Vitamin A, Vitamin K, Arsenic, Boron, Chromium, Copper, Iodine, Iron, Manganese, Molybdenum, Nickel, Silicon, Vanadium, and Zinc*; Food and Nutritional Board: Washington, DC, USA, 2001.

38. Anderson, R.A.; Bryden, N.A.; Polansky, M.M. Dietary chromium intake. *Biol. Trace Elem. Res.* **1992**, *32*, 117–121. [CrossRef] [PubMed]

39. European Food Safety Authority. *Tolerable Upper Intake Levels for Vitamins and Minerals*; European Food Safety Authority: Parma, Italy, 2006.

40. Scientific Committee for Food. *Reports of the Scientific Committee for Food of the European Community (Thirty-First Series)*; Nutrient and Energy Intakes for the European Community; Scientific Committee for Food: Luxembourg, 1993.

41. World Health Organization. *Worldwide Prevalence of Anaemia 1993–2005—WHO Global Database on Anaemia*; World Health Organization: Geneva, Switzerland, 2008.

42. USDA National Nutrient Database for Standard Reference. Available online: https://ndb.nal.usda.gov/ndb/search/list?SYNCHRONIZER_TOKEN=c7ce3c7c-2729-4297-9d40-68b5e1b99a2b&SYNCHRONIZER_URI=%2Fndb%2Fsearch%2Flist&qt=&qlookup=&ds=SR&manu= (accessed on 10 May 2018).

43. Hambridge, K.M.; Casey, C.E.; Krebs, N. Zinc. In *Trace Elements in Human and Animal Nutrition*; Merts, W., Ed.; Academic Press Inc: Orlando, FL, USA, 1986; pp. 1–137.

agriculture

MDPI

Article

Phytochemical Analysis and Antioxidant Properties in Colored Tiggiano Carrots

Aurelia Scarano [1], Carmela Gerardi [1], Leone D'Amico [1], Rita Accogli [2] and Angelo Santino [1,*]

[1] C.N.R. Unit of Lecce, ISPA-CNR, Institute of Science of Food Production, 73100 Lecce, Italy; aurelia.scarano@ispa.cnr.it (A.S.); carmela.gerardi@ispa.cnr.it (C.G.); leone.damico@ispa.cnr.it (L.D.)
[2] Orto Botanico Dipartimento di Scienze e Tecnologie Biologiche ed Ambientali, Università del Salento, 73100 Lecce, Italy; rita.accogli@unisalento.it
* Correspondence: angelo.santino@ispa.cnr.it; Tel.: +39-0832422606

Received: 31 May 2018; Accepted: 20 June 2018; Published: 2 July 2018

Abstract: The carrot (*Daucus carota* L.) is an important vegetable source of bioactive compounds in the human diet. In the Apulia region (Southern Italy), local farmers have domesticated colored landraces of carrots over the years, strictly related to local cults and traditions. Amongst these, an important landrace is the carrot of Saint Ippazio or the Tiggiano carrot. In the present study, we evaluated the content of carotenoids, anthocyanins, phenolic acids, sugars, organic acids, and antioxidant activity in Tiggiano carrots. Our results indicated that yellow-purple carrots have the highest levels of bioactive compounds, together with the highest antioxidant capacity compared to the yellow and cultivated orange varieties. These data point out the nutritional value of purple Tiggiano carrots and may contribute to the valorization of this typical landrace.

Keywords: apulian landraces; bioactive compounds; polyphenols; Tiggiano carrot

1. Introduction

The carrot is a root vegetable widely consumed in human diet, either as fresh or processed in meals and beverages. The carrot have been ranked 10th among 39 fruits and vegetables for its multiple nutritional benefits [1]. Italy accounts for over 500,000 tons of production [2], with about 300,000 quintals coming from the Apulia region in 2017 [3]. Carrots are an important dietary source of carotenoids [4], mostly α-carotene and β-carotene, also known as provitamin A, since it can be converted to vitamin A once in the body. Despite western countries consuming commonly orange carrots, yellow or purple colored carrots are also cultivated in some parts of the U.S. [5], Europe, Turkey, and India [6]. The domestication of yellow and purple carrots arises from the oriental countries spreading to Italy starting from the 13–14th century [7,8]. In the Apulia region, a plan of preserving crop biodiversity has been launched from the Regional Administration (2007–2013 Rural Development Program, actions 214/3 and 214/4), with the aim of rescuing and valorizing the local landraces of crops traditionally cultivated in the rural areas. Several Apulian varieties are included in the list of vegetables at risk of genetic erosion [9], which constitutes an issue for the loss of the genetic traits important for the biodiversity and nutritional quality of the varieties [8–10]. In the Apulia region, three main ecotypes of colored carrots have been reported, whose names are strictly related to the production areas: Polignano (province of Bari, [11]), Zapponeta, (province of Foggia), and Tiggiano (which also includes Tricase and Specchia villages, in the province of Lecce [12]). Tiggiano carrot cultivation is strongly related to the popular cult of the Saint Ippazio, protector of male virility. In celebration of the saint, the colored carrots are usually sold as a pagan ritual concerning fertility, and to receive protection from the saint against hernias or male impotency. As in the case of Polignano carrot [8], one of the phenotypes of the Tiggiano carrot has a dark purple epidermis and a yellow-orange inner core. In many varieties of purple carrots, the dark color has been attributed to the accumulation of anthocyanins in the taproots [5,13–16].

Anthocyanins are natural pigments widely occurring in plants, which contribute to the nutritional value of vegetable and fruits, due to their molecular antioxidant properties and their involvement in anti-aging and anti-inflammatory processes [17–19]. In plants, anthocyanins are synthetized by the flavonoid biosynthetic pathway, which leads also to the production of phenolic acids and other classes of polyphenols with healthy benefits [20–22]. In this study, we carried out the chemical profiling of antioxidant compounds and characterized the main classes of polyphenols, carotenoids, sugars, and organic acids, which can contribute to the quality and nutritional value of Tiggiano carrots and to the valorization of this local landrace.

2. Materials and Methods

2.1. Plant Material

The carrots were grown and harvested in a local farm in Tiggiano (Lecce), South Italy. Samples were divided in three groups, based on the color of the epidermis of the carrots (yellow, orange, or purple). Five carrots for each group were used for the analyses. Carrots were cut, frozen in liquid nitrogen, grinded in a mortar, and stored at -80 °C for anthocyanin analyses. Alternatively, for the determination of phenolic acids, carotenoids, organic acids, sugars, and antioxidant activity, samples were lyophilized, finely powdered, and stored at 4 °C until their use. All analyses were performed in triplicate.

2.2. Extraction and Detection of Phenolic Acids

Two-hundred mg of freeze-dried powder of samples were extracted three times in methanol: water 80:20 (v/v), the extracts centrifuged, and the supernatants combined. Phenolic acids were detected at 320 nm by RP-HPLC DAD (Agilent 1100 HPLC system, Agilent Technologies Inc., Santa Clara, CA, USA). Separation was performed on a C18 column (5 UltraSphere, 80 A pore, 25 mm), with a linear gradient from 20% to 60% acetonitrile, in 55 min, with a flow of 1 mL/min at 25 °C. Concentrations were obtained by referring to calibration curves and results were expressed in µg/g dried weight.

2.3. Total Content of Anthocyanins

The total content of anthocyanins was determined on methanol extracts from fresh Tiggiano carrots using the pH differential method [23]. Fresh samples were frozen in liquid nitrogen and ground in a mortar. Two-hundred mg powdered samples were extracted three times in methanol: water 80:20 (v/v). The methanol extracts were mixed using the appropriate dilution factor, with two different solutions to obtain different pH values, prepared as previously described [23]: pH 1.0 potassium chloride buffer (0.025 M KCl) and pH 4.5 sodium acetate buffer (0.4 M $CH_3CO_2Na \cdot 3H_2O$). After 15 min incubation at room temperature, the absorbance of the samples was measured at 520 nm and 700 nm (Shimadzu UV-1800, spectrophotometer, Kyoto, Japan). The total content of anthocyanins, expressed as cyanidin-3-glucosyde equivalents, was calculated according to the formula described in Lee et al. (2005) [23].

2.4. Sugar and Organic Acid Extraction and Quantification

Sugars and organic acids were extracted two times by mixing 200 mg of freeze-dried powder with 200 mg of PVPP in 10 mL of Milli-Q-water for 1 hour at room temperature. After centrifugation of the slurry (10 min at 4000× g), the supernatants were collected and 1 mL of extract was further centrifuged (10 min at 15,000× g and injected into the HPLC system (Agilent 1100 series)). The identification and quantification were performed by the HPLC system equipped with a pump system, a refractive index detector (RID) for sugar analysis, and a UV/Vis detector monitoring organic acids at 210 nm onto an Aminex HPX-87H column (300 × 7.8, 9 μm) (Bio-Rad, Hercules, CA, USA), kept at 55 °C. The analytical method was the same as reported in Gerardi et al. (2015) [24].

2.5. Carotenoid Content

Samples were frozen in liquid nitrogen, grinded in a mortar until finely powdered, and stored at −80 °C until the analysis. Extractions were performed according to the method described in Koch and Holdman (2004) [25]. The supernatants were combined and dried with a rotary evaporator. Samples were resuspended in 1 ml of ethyl acetate and analyzed by RP-HPLC DAD as previously described [22].

2.6. Determination of Antioxidant Activity

The TEAC assay was performed as previously reported [26]. Briefly, 2,2′-azinobis(3-ethylbenzothiazoline-6-sulfonic acid) diammonium salt (ABTS, Sigma-Aldrich, St. Louis, MO, USA) radical cations were prepared by mixing an aqueous solution of 2.45 mM potassium persulfate (final concentration) and an aqueous solution of 7 mM ABTS (final concentration) and were allowed to stand in the dark at room temperature for 12–16 h, before use. The ABTS radical cation solution was diluted in PBS (pH 7.4) to an absorbance of 0.40 at 734 nm. Trolox was used as antioxidant standard and to prepare a standard calibration curve (0–16 μM). After the addition of 200 μL of diluted ABTS to 10 μL of Trolox standard or extracts diluted in PBS, in each well of a 96 well-plate (Costar), the absorbance reading at 734 nm was taken 6 min after initial mixing using an Infinite200Pro plate reader (Tecan, Männedorf, Swizerland). Appropriate solvent blanks were run in each plate. All extracts were assayed at least at three separate dilutions and in triplicate. The percentage inhibition of absorbance at 734 nm is calculated and plotted as a function of concentration of Trolox and the TEAC value expressed as Trolox equivalent (in μmolar) using Magellan v7.2 software.

2.7. Statistical Analysis

Values were expressed as mean ± SD of three independent experiments. One-way analysis of variance (ANOVA) was performed, followed by separation of means with Tukey's multiple comparison test, using GraphPad Prism version 5.0 (GraphPad Software, La Jolla, CA, USA).

3. Results

Tiggiano carrots are characterized by different colors of epidermis. Figure 1a shows representative yellow, orange and yellow-purple carrots analyzed in this study. It is worth mentioning that sections of the purple carrots showed that this color was mainly limited to the epidermis, whereas the cortex and the vascular cylinder showed a typical purple-orange and yellow-pigmented color, respectively (Figure 1b).

Due to their colors, we firstly investigated the presence of anthocyanins (Figure 2a). The anthocyanins content was over 100 mg cyanidin-3-glucoside equivalents/100 g of fresh weight in Tiggiano yellow-purple carrots. As expected, anthocyanins were undetected in the yellow and orange carrots. We also analyzed the content of phenolic acid, another phytochemical belonging to the group of polyphenols. As shown in Figure 2b, the phenolic acid content in the Tiggiano carrots was higher compared to the orange commercial cultivar used as control. Among the Tiggiano carrots, the yellow-purple ones showed the highest amount of these polyphenols, with chlorogenic acid as the main representative (−2.6 mg/g dry weight, compared to −1 mg/g dry weight detected in the commercial carrot cultivar), followed by caffeic acid (−0.26 mg/g DW compared to −0.18 mg/g DW in the commercial carrot cultivar). Other phenolic acids, mainly *p*-coumaric acid and ferulic acid, were detected in traces.

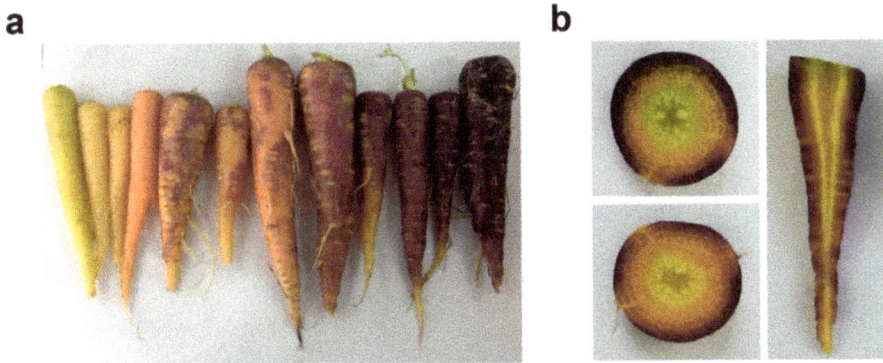

Figure 1. (**a**) Multicolored phenotypes of Tiggiano carrots. (**b**) Sections of a purple Tiggiano taproot in detail. The epidermis shows the typical purple color followed by the cortex (purple-orange pigmented) and the inner core, represented by the vascular cylinder (yellow pigmented).

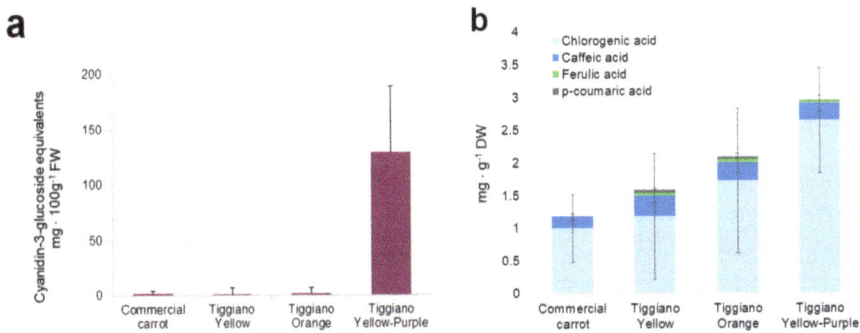

Figure 2. (**a**) Anthocyanins and (**b**) phenolic acids content in Tiggiano carrots compared to a commercial cultivar. Data are expressed as means ($n = 3$) $\pm$ SD.

We investigated the content of organic acids and sugars, which are an important quality parameter and are involved in the taste and flavor of vegetables, influencing the overall level of acidity of the fruits. Three main organic acids were identified in Tiggiano carrots: malic, tartaric, and citric acids. As shown in Figure 3a, malic acid was the most abundant, both in commercial and Tiggiano carrots. Malic acid content in the Tiggiano carrots was higher than in the commercial cultivar but this result was not statistically significant. On the other hand, the level of citric acid was significantly lower in Tiggiano carrots than in the commercial orange cultivar.

Using the method described in Materials and Methods, we were able to obtain a simultaneous identification and quantification of organic acid and monosaccharides. Concerning the monosaccharides, the most abundant was glucose followed by fructose, where either yellow or orange Tiggiano carrots showed a significant higher glucose content (0.5 g/g of dry weight) compared to the commercial carrot and purple/orange Tiggiano carrot.

Next, we analyzed the carotenoid content in the different colored Tiggiano carrots. As illustrated in Figure 4, β-carotene was the main carotenoid found in our analyses, reaching the highest levels in the commercial and yellow-purple carrots. Significantly low levels were recorded in yellow/orange Tiggiano carrots. Concerning lutein, a significantly higher level was observed in yellow-purple Tiggiano carrots (−90 μg/g DW) compared to the commercial or yellow/orange Tiggiano carrots.

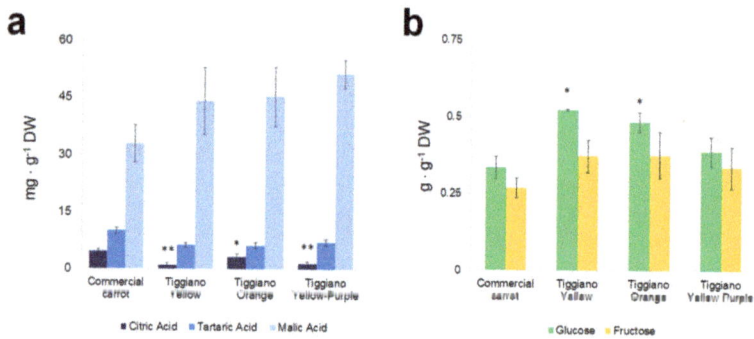

Figure 3. (**a**) Main organic acids and (**b**) sugars detected in Tiggiano carrots compared to the commercial cultivar. Data are expressed as mean ± SD. Significance assumed at * $p < 0.05$, ** $p < 0.01$.

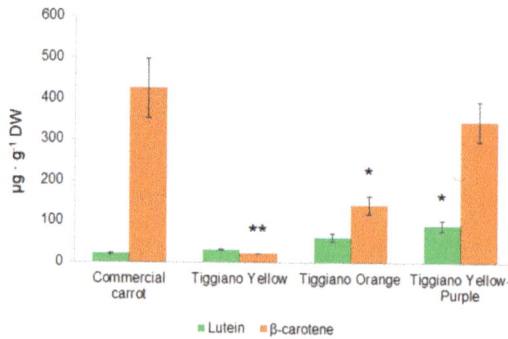

Figure 4. Carotenoid content in Tiggiano carrots compared to a commercial cultivar. Significance assumed at * $p < 0.05$, ** $p < 0.01$.

Finally, we measured the antioxidant activity in Tiggiano carrots by the Trolox Equivalent Antioxidant Capacity (Figure 5). Our results pointed out a similar anti-oxidant capability in the extract of yellow or orange Tiggiano carrots and the commercial carrot. Notably, the extracts from purple carrots showed a significant higher antioxidant capacity in the hydrophilic fraction, which doubled that of the commercial carrot and those from the yellow/orange Tiggiano carrots.

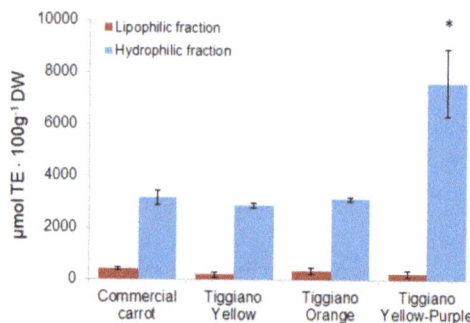

Figure 5. Antioxidant profile of Tiggiano carrots compared to a commercial cultivar. Significance assumed at * $p < 0.05$.

4. Discussion

In this study, we carried out a phytochemical and antioxidant profiling of the main components associated with the quality and nutritional value of the Apulian colored Tiggiano carrots. Yellow and orange Tiggiano carrots showed different features in respect to the yellow-purple carrots. Compared to the yellow-purple ones, the yellow and orange Tiggiano carrots showed higher glucose and lower citric acid levels. Both these parameters influence the organoleptic properties of food products [27], since sugars reduce the perception of acidity, affecting the taste and the sweetness of fruits [7,28]. On the other hand, the reduced levels of citric acid in all Tiggiano carrots may influence the astringency sensation [29].

Notably, the yellow-purple Tiggiano carrots also showed a similar β-carotene content, but higher lutein levels compared to the commercial cultivar or yellow/orange Tiggiano carrots, a result that can further contribute to the nutritional quality of these colored fruits.

Moving to other phytochemicals related to the nutritional value of fruit crops, our results clearly indicated that high levels of phenolic acids (such as caffeic acid and chlorogenic acid) and anthocyanins were specifically associated with yellow-purple carrots. All these compounds have been associated to the health-promoting properties of fruits and vegetables [19,22]. Our results are in agreement with other studies, which reported a higher content of these phytochemicals in the cortex of different purple carrot varieties [5].

Tiggiano carrots show similarities with Polignano carrots, another Apulian landrace previously characterized [8]. Similarly to the Tiggiano yellow-purple carrots, Polignano carrots are also characterized by a high content of phenolic acids and polyphenols assayed by Folin-Ciocalteau assay [8]. Furthermore, Polignano carrots showed a higher glucose and fructose content than the commercial carrots [8] confirming our results on carbohydrate content of Tiggiano carrots. Taken together, these results revealed interesting similarities in these two landraces. Further studies are now in progress to verify the genetic relationships between Tiggiano and Polignano carrots.

Anthocyanins are natural pigments with a protective function against abiotic (UV irradiation, cold) or biotic stresses. However, the reasons why the anthocyanin biosynthetic pathway is activated in yellow-purple and not in orange or yellow carrots are still unknown. In Tiggiano yellow-purple carrots, we also found a higher antioxidant capacity compared to the other carrots, which likely contributes to increasing the nutritional quality of the fresh product. On these bases, their consumption might have positive effects on human health. Other studies indicate that diets rich in anthocyanins and phenolic acids have protective effects against oxidative stress, involved in the onset of aging processes and several human pathologies [21,22].

5. Conclusions

Crop landrace conservation is an emerging challenge to be addressed, due to progressive susceptibility to the genetic erosion of local crops, as in the case of the Apulian carrot varieties. In general, the cultivation of landraces is decreasing because of the modern cropping systems that use hybrids with features mainly satisfying the requests of the global market. Nevertheless, some local varieties still maintain genetic traits adapted to be more efficient in stress defense and nutritive uptake and, in some cases, they are able to accumulate phytonutrients, thus showing interesting features for the human diet. In the local carrot landraces analyzed in this study, besides their role in nutritional quality, the presence of bioactive compounds highlights on the possible activation of the anthocyanin biosynthetic pathway in the taproots. Therefore, further studies will be useful to elucidate the genetic features of the different Apulian carrots landraces. Despite the fact that it is still unclear if the Apulian colored carrots derive from the same genetic background, they represent important local genetic resources, to be preserved and valorized.

Author Contributions: A.S. (Aurelia Scarano), L.D., C.G. performed the experiments; A.S. (Aurelia Scarano), L.D., C.G., A.S. (Angelo Santino) analyzed the data; A.S. (Aurelia Scarano), R.A., A.S. (Angelo Santino) wrote the paper with the approval of all the authors.

Funding: This work was supported by the Apulia Region project "BiodiverSO".

Conflicts of Interest: The authors declare no conflict of interest.

References

1. Silva Dias, J.C. Nutritional and health benefits of carrots and their seed extracts. *Food Nutr. Sci.* **2014**, *5*, 2147–2156. [CrossRef]
2. Food and Agriculture Organization of the United Nations. *FAOSTAT Statistic Database*; Food and Agriculture Organization of the United Nations: Rome, Italy, 2017.
3. ISTAT. Italian National Institute of Statistics, Rome. 2017. Available online: http://agri.istat.it/ (accessed on 1 February 2018).
4. Sharma, K.D.; Karki, S.; Thakur, N.S.; Attri, S. Chemical composition, functional properties and processing of carrot—A review. *J. Food Sci. Technol.* **2012**, *49*, 22–32. [CrossRef] [PubMed]
5. Sun, T.; Simon, P.W.; Tanumihardjo, S.A. Antioxidant phytochemicals and antioxidant capacity of biofortified carrots (*Daucus carota* L.) of various colors. *J. Agric. Food Chem.* **2009**, *57*, 4142. [CrossRef] [PubMed]
6. Arscott, S.A.; Tanumihardjo, S.A. Carrots of many colors provide basic nutrition and bioavailable phytochemicals acting as a functional food. *Comp. Rev. Food Sci. Food Saf.* **2010**, *9*, 223. [CrossRef]
7. Simon, P.W. Domestication, historical development, and modern breeding of carrot. *Plant Breed. Rev.* **2000**, *19*, 157–190.
8. Cefola, M.; Pace, B.; Renna, M.; Santamaria, P.; Signore, A.; Serio, F. Compositional analysis and antioxidant profile of yellow, orange and purple Polignano carrots. *Ital. J. Food Sci.* **2012**, *24*, 284–291.
9. Renna, M.; Serio, F.; Signore, A.; Santamaria, P. The yellow-purple Polignano carrot (*Daucus carota* L.): A multicolored landrace from the Puglia region (Southern Italy) at risk of genetic erosion. *Gen. Res. Crop Evol.* **2014**, *61*, 1611–1619. [CrossRef]
10. Elia, A.; Santamaria, P. Biodiversity in vegetable crops, a heritage to save: The case of Puglia region. *Ital. J. Agron.* **2013**, *8*, 4. [CrossRef]
11. Signore, A.; Renna, M.; D'Imperio, M.; Serio, F.; Santamaria, P. Preliminary evidences of biofortification with iodine of "Carota di Polignano", an Italian carrot landrace. *Front. Plant Sci.* **2018**, *9*, 170. [CrossRef] [PubMed]
12. Accogli, R.; Marchiori, S. Verifica agronomica di *Daucus carota* L. varietà locale carota de Santu Pati. In *Progetto Co.Al.Ta. 1 "Analisi e Valutazione di Ordinamenti Produttivi Alternativi Nelle Aree a Riconversione del Tabacco" Risultati del 1° Anno di Attività*; A Cura dell'Istituto INEA: Rome, Italy, 2006.
13. Cuevas Montilla, E.; Rodriguez Arzaba, M.; Hillebrand, S.; Winterhalter, P. Anthocyanin composition of black carrot (*Daucus carota* ssp. *sativus* var. *atrorubens* Alef.) cultivars Antonina, Beta Sweet, Deep Purple, and Purple Haze. *J. Agric. Food Chem.* **2011**, *59*, 3385–3390.
14. Yildiz, M.; Willis, D.K; Cavagnaro, P.F.; Iorizzo, M.; Abak, K.; Simon, P.W. Expression and mapping of anthocyanin biosynthesis genes in carrot. *Theor. Appl. Genet.* **2013**, *126*, 1689–1702. [CrossRef] [PubMed]
15. Xu, Z.-S.; Huang, Y.; Wang, F.; Song, X.; Wang, G.-L.; Xiong, A.-S. Transcript profiling of structural genes involved in cyanidin-based anthocyanin biosynthesis between purple and non-purple carrot (*Daucus carota* L.) cultivars reveals distinct patterns. *BMC Plant Biol.* **2014**, *14*, 262. [CrossRef] [PubMed]
16. Xu, Z.-S.; Feng, K.; Que, F.; Wang, F.; Xiong, A.-S. A MYB transcription factor, DcMYB6, is involved in regulating anthocyanin biosynthesis in purple carrot taproots. *Sci. Rep.* **2017**, *7*, 45324. [CrossRef] [PubMed]
17. Butelli, E.; Titta, L.; Giorgio, M.; Mock, H.P.; Peterek, S.; Schijlen, E.G.; Hall, R.D.; Bovy, A.G.; Luo, J.; Martin, C. Enrichment of tomato fruit with health-promoting anthocyanins by expression of selected transcription factors. *Nat. Biotechnol.* **2008**, *26*, 1301–1308. [CrossRef] [PubMed]
18. Tomlinson, M.L.; Butelli, E.; Martin, C.; Carding, S.R. Flavonoids from engineered tomatoes inhibit gut barrier pro-inflammatory cytokines and chemokines, via SAPK/JNK and p38 MAPK pathways. *Front. Nutr.* **2017**, *4*, 61. [CrossRef] [PubMed]

19. Blando, F.; Calabriso, N.; Berland, H.; Maiorano, G.; Gerardi, C.; Carluccio, M.A.; Andersen, Ø.M. Radical scavenging and anti-inflammatory activities of representative anthocyanin groupings from pigment-rich fruits and vegetables. *Int. J. Mol. Sci.* **2018**, *19*, 169. [CrossRef] [PubMed]

20. Galleggiante, V.; De Santis, S.; Cavalcanti, E.; Scarano, A.; De Benedictis, M.; Serino, G.; Caruso, M.L.; Mastronardi, M.; Pinto, A.; Campiglia, P.; et al. Dendritic cells modulate iron homeostasis and inflammatory abilities following quercetin exposure. *Curr. Pharm. Des.* **2017**, *23*, 2139–2146. [CrossRef] [PubMed]

21. Santino, A.; Scarano, A.; De Santis, S.; De Benedictis, M.; Giovinazzo, G.; Chieppa, M. Gut microbiota modulation and anti-inflammatory properties of dietary polyphenols in IBD: New and consolidated perspectives. *Curr. Pharm. Des.* **2017**, *23*, 2344–2351. [CrossRef] [PubMed]

22. Scarano, A.; Butelli, E.; De Santis, S.; Cavalcanti, E.; Hill, L.; De Angelis, M.; Giovinazzo, G.; Chieppa, M.; Martin, C.; Santino, A. Combined dietary anthocyanins, flavonols, and stilbenoids alleviate inflammatory bowel disease symptoms in mice. *Front. Nutr.* **2018**, *4*, 75. [CrossRef] [PubMed]

23. Lee, J.; Durst, R.W.; Wrolstad, R.E. Determination of total monomeric anthocyanin pigment content of fruit juices, beverages, natural colorants, and wines by the pH differential method: Collaborative study. *J. AOC Int.* **2005**, *88*, 1269–1278.

24. Gerardi, C.; Tommasi, N.; Albano, C.; Pinthus, E.; Rescio, L.; Blando, F.; Mita, G. *Prunus mahaleb* L. fruit extracts: A novel source for natural pigments. *Eur. Food Res. Technol.* **2015**, *241*, 683–695. [CrossRef]

25. Koch, T.; Goldman, I. A one-pass semi-quantitative method for extraction and analysis of carotenoids and tocopherols in carrot. *HortScience* **2004**, *39*, 1260–1261.

26. Prior, R.L.; Wu, X.; Schaich, K. Standardized methods for the determination of antioxidant capacity and phenolics in foods and dietary supplements. *J. Agric. Food Chem.* **2012**, *53*, 4290–4302. [CrossRef] [PubMed]

27. Mikulic-Petkovsek, M.; Schmitzer, V.; Slatnar, A.; Stampar, F.; Veberic, R. Composition of sugars, organic acids, and total phenolics in 25 wild or cultivated berry species. *J. Food Sci.* **2012**, *77*, C1064–C1070. [CrossRef] [PubMed]

28. Bae, H.; Yun, S.K.; Jun, J.H.; Yoon, I.K.; Nam, E.Y.; Kwon, J.H. Assessment of organic acid and sugar composition in apricot, plumcot, plum, and peach during fruit development. *J. Appl. Bot. Food. Qual.* **2014**, *87*, 24–29.

29. Bajec, M.R.; Pickering, G.J. Astringency: Mechanisms and Perception. *Crit. Rev. Food Sci. Nut.* **2008**, *48*, 858–875. [CrossRef] [PubMed]

agriculture

MDPI

Article

Conservation of Crop Genetic Resources in Italy with a Focus on Vegetables and a Case Study of a Neglected Race of Brassica Oleracea

Karl Hammer [1], Vincenzo Montesano [2], Paolo Direnzo [2] and Gaetano Laghetti [2,*]

[1] Leibnitz-Institut für Pflanzengenetik und Kulturpflanzenforschung (IPK), Corrensstr. 3, OT Gatersleben, D-06455 Seeland, Germany; khammer.gat@t online.de
[2] National Research Council-Institute of Biosciences and BioResources (CNR-IBBR), via Amendola 165/A, 70126 Bari, Italy; vincenzo.montesano@ibbr.cnr.it (V.M.); paolo.direnzo@ibbr.cnr.it (P.D.)
*   Correspondence: gaetano.laghetti@ibbr.cnr.it; Tel.: +39-080-5583400

Received: 3 May 2018; Accepted: 26 June 2018; Published: 2 July 2018

Abstract: This study attempts, above all, to provide a summary, on a strictly scientific basis, about the strategies of conservation of autochthonous agrobiodiversity followed in Italy. A special focus is dedicated to vegetables and, therefore, could represent a contribution to improve the national strategy for the safeguarding of its agrobiodiversity in general. The paper offers also an outlook on the most critical factors of ex situ conservation and actions which need to be taken. Some examples of 'novel' recovered neglected crops are also given. Finally a case study is proposed on 'Mugnolicchio', a neglected race of *Brassica oleracea* L., cultivated in Altamura (Ba) in southern Italy, that might be considered as an early step in the evolution of broccoli (*B. oleracea* L. var. italica Plenck) like 'Mugnoli' another neglected race described from Salento (Apulia).

Keywords: agrobiodiversity; vegetables; plant genetic resources; Italy; safeguarding; landraces

1. Introduction

The present study is a small review reported by the authors after a number of safeguarding projects and collecting missions had been carried out, since the 1970s, in all Italian agricultural districts [1,2], including small islands [3] and linguistic areas [4].

1.1. Conservation of Crop Genetic Resources: The Italian Situation

To understand the role and the importance of agrobiodiversity in the Italian agricultural system, it is interesting to know the statistics that describes it: one has the impression of standing in front of a country still caught between tradition and modernity, where agricultural activities—today an insignificant percentage of GDP—still retain their value for a large part of the population. In fact, despite the decline in recent years, Italy is the third largest agricultural country in Europe after Poland and Romania, with more than a million employees in the sector. Also for the number of companies in agriculture, Italy holds third place, again after Romania and Poland. In this framework, agrobiodiversity plays a dual role: on the one hand, it is still strongly linked with farmers who manage their farms traditionally and not as real "enterprises" and, on the other hand, their highly qualitative production awarded by many geographical indications, e.g., Protected Designation of Origin (PDO) as the scarlet eggplant (*Solanum aethiopicum* L.), "Melanzana rossa del Pollino", as the common bean "Fagiolo poverello bianco di Rotonda", Protected Geographical Indication (PGI) as the common beans "Fagioli di Sarconi", and Traditional Specialities Guaranteed (TSG) represent worldwide excellence. Italy, for the latter, is the queen of Europe with more than 200 certified products, which represent more than 20% of the European total. "Geographical indication" trademarks are a demonstration of the link between territory, culture, and agriculture; their strong presence in Italy attests the importance which this trio still has in shaping the economic development of Italian agriculture. It should be noted, however, that most of the agrobiodiversity and traditional knowledge associated with it, is kept in a class of farms generally conducted by elder farmers over 65 years of age [1,2,5].

It is necessary, therefore, to adopt policies to cope with this situation, and to avoid loss of knowledge and of landraces due to generational change, and to create economic, social, and cultural conditions for these farms to continue working in agriculture. In fact, the market and international competition are horizons too far away from them that, without adequate forms of protection or development, would disappear, taking with them all the specific culture handed down from generation to generation. In this context, agricultural policies play a central role, in particular, those of rural development, which can, if properly set up, promote the link between tradition and modernity, avoiding interruptions and using agrobiodiversity as a factor in local development. For this reason, it is not only a simple implementation of conservation policies for plant genetic resources, but also a change of perspective by moving towards a system of safeguarding to provide a reciprocal interaction and a necessary complementary action between ex situ and in situ/on-farm conservation.

The Regions and the Autonomous Provinces are public bodies which, by their deep knowledge of the territory and their legislative autonomy in the field of agriculture, are the privileged places to synthesize and coordinate the main actions of conservation and exploitation of agrobiodiversity. In fact, there are many regions that fund and promote in various ways such actions in their territories. In some cases, these activities have led to specific regional legislation with the aim of protecting local breeds and varieties. Tuscany was the first region to enact a law on the protection of agrobiodiversity in 1997, followed in subsequent years by Lazio, Umbria, Friuli Venezia-Giulia, Marche, Emilia-Romagna, Basilicata, and Apulia. At present also other regions are discussing to enact similar laws [5]. The experiences of Italian regional laws can be considered as one of the few operative examples in Europe for protection and collection of Plant Genetic Resources (PGR). They have anticipated policies at national and European level, even if operating in line with the objectives of the principles of the Convention on Biodiversity, failing to implement the simplifications ensured by the Multilateral Systems of the International Treaty (such as most importantly the use of a standard material transfer agreement with standard terms of access and benefit sharing). In Italy, however, in addition to the regions, there are several entities, variously integrated with each other, depending on the territorial dynamics, that interact towards building a chain of plant genetic resources, from storage to exploitation. There are three categories of entities: scientific institutions, local authorities, and the non-governmental sector. The three categories should work in a completely synergistic way with each other. In general, these are:

- Scientific institutions dealing with collecting, preservation, characterization and documentation of material and ex situ conservation, as well as dissemination of the information collected;
- Regions, Autonomous Provinces and other local institutions (Provinces, Municipalities, Mountain Communities, GAL—Groups of Local Action, etc.) coordinate and promote these actions often supporting them with dedicated lines of credit (e.g., regional laws for the protection of agrobiodiversity) or through funds for agricultural regional research and the "Plans of Rural Development" or others;
- The non-governmental sector (all subjects not included in the previous two categories, such as individually or jointly working farmers, associations, foundations, various organizations, etc.) stimulates and/or carries out paths of preservation and exploitation of specific landraces or particular territories, starting from the needs of local communities and farmers and their history.

In this context, the role of farmers is crucial. They are important both as farmers as such (growing landraces in their farms), as "guardian farmers", and as associate members in programs to exploit and promote specific PGR.

Consumers are also particularly interested in landraces, so that a vibrant market for local and/or typical products is created. Typicity presumes that a local variety, its product, and any process of transformation are closely linked to the territory in which the genetic resource has evolved. The term "territory" should be used in the broadest and most complete sense, indicating both the physical space and anthropological space (typical elements of the mode of man settlement), as well as the set of values, history and culture that characterize it.

In recent years, there have been many experiences of conservation and exploitation of landraces by private persons (farmers and non-farmers) who autonomously have provided funds for projects often linked to the promotion of a particular territory and products connected to it. These initiatives are dispersed throughout the country (through, e.g., fairs, markets, dissemination, promotion and exploitation actions, consortia of producers, development of product rules, small projects on typical products), which over time have shown a strong fragmentation, poor coordination, and frequent overlap, but most have failed to transmit adequately the "know-how". It must be said, however, that the dissemination activities, including publications produced in recent years, have contributed in a concrete way to the knowledge of the heritage of Italian landraces, which often did not find adequate description in the official manuals (e.g., scientific journals, descriptive sheets of Guide Lines by INEA [6]). The collection of information derived from cookbooks and popular knowledge should not be underestimated, which allows proper cultivation and use of old landraces. The wealth of material and knowledge created in the past from ancient and disinterested experience of farmers is a precious inheritance that has to remain "World Heritage".

1.2. Plant Genetic Resources Stored by Italian Public Institutions and Universities

The depletion of PGR has important implications both ecologically and economically. The erosion and possible extinction of these resources can undermine the resilience of ecosystems and endanger the essential environmental services derived from them. For the economy, PGR are a source of direct and indirect benefits. They are indeed a source of raw materials as well as useful information, for example, in the processes of plant breeding of crops. The Mediterranean, and particularly its less developed rural areas, is traditionally rich in PGR which, however, are undergoing a process of genetic erosion due to causes both socio-economic, such as the marginalization of agriculture, and environmental, as in the case of the loss of natural habitats [7].

The Italian national activities of inventorying PGR for food and agriculture, promoting the collecting and safeguarding, to establish a network of updated information on PGR, are concentrated mainly in the "Council for Research and Experimentation in Agriculture and Agricultural Economic Analysis" (CREA, [8]) and the National Research Council (CNR, [9]).

Although it is known that many universities maintain large collections of agricultural genetic resources, a comprehensive list has never been compiled. Several universities store remarkable collections and work in areas rich in crop diversity. The Department of Applied Biology (University of Perugia), for example, has important collections of forage species (legumes and grasses), food, as well as industrial, medicinal and aromatic crops while the Centre for Conservation and Exploitation of Plant Biodiversity (University of Sassari) has collections of seed germplasm and DNA of populations of native endemic species of high phyto-geographical interest, collections of cultivars of fruit and vegetables, and micro-organisms—both pathogens and symbionts. In Sicily, instead, a specific measure of a regional law (POR 2000–2006) allowed the Universities of Palermo and Catania, the CREA and the CNR, to create several centers for the in vivo and in vitro conservation of germplasm of fruit trees, olive and citrus that could be networked together, sharing information and contributing to the knowledge on all plant material in storage [10,11].

1.3. PGR Stored by the Research Institutes of the CREA

The MiPAAF (Italian Ministry of Agriculture), to deal with these and other international commitments, financed in 1999 and 2001 two nationwide projects aimed at a census of PGR for agriculture preserved ex situ at the Institutes for Experimental Research in Agriculture (former IRSA, now institutes of CREA) and the fruit germplasm preserved ex situ in various Italian institutions of different backgrounds (IRSA, CNR, universities, regional experimental farms). Since 1995, the focal point of coordination actions on PGR is the CREA-FRU (Institution acronyms are explained in Table 1 below.), which, over the years, has established itself as the reference point for the MiPAAF both nationally and internationally with regard to the PGR.

Table 1. Plant genetic resources of agricultural interest and research units of the project "Plant Genetic Resources/Implementation of the FAO Treaty" (modified from [12]).

Plant Genetic Resources	CREA Institutes	No. Accessions (CREA)	CNR Institutes	No. Accessions (CNR)	Other Research Units	No. Accessions ("Semi Rurali")
cereals	ACM, CER, GPG, MAC, QCE, RIS, SCV	17,496	IBBR	34,920	"Semi Rurali" Network	1190 (total, mostly cereals)
vegetables	ORA, ORL, ORT	–	IBBR, ISAFOM	3844	"Semi Rurali" Network	
fruits and nuts	ACM, FRC, FRF, FRU, SCA	8787	IVALSA	6160	"Semi Rurali" Network	
fodder species	FLC	7776	IBBR	6561	–	–
Industrial crops	API, CAT, CIN	2714	–	–	–	–
olive	OLI	3243	IVALSA, ISAFOM, IBBR	2500	–	–
grape	VIT	793	IBBR	119	–	–
ornamental species	FSO, SFM, VIV	266	–	–	–	–
medicinal and aromatic plants	MPF	586	IBBR	448	–	–
forest species	SEL, PLF	3.744	IBBR	5.326	–	–

ACM (Centro di ricerca per l'agrumicoltura e le colture mediterranee, Acireale), API (Unità di ricerca di apicoltura e bachicoltura, Bologna), CAT (Unità di ricerca per le colture alternative al tabacco, Scafati), CER (Centro di ricerca per la cerealicoltura, Foggia), CIN (Centro di ricerca per le colture industriali, Bologna e Rovigo), FLC (Centro di Ricerca per le Produzioni foraggere e lattiero-casearie, Lodi), FRC (Unità di ricerca per la frutticoltura, Caserta), FRF (Unità di ricerca per la frutticoltura, Forlì), FRU (Centro di ricerca per la frutticoltura, Roma), FSO (Unità di ricerca per la floricoltura e le specie ornamentali, Sanremo), GPG (Unità di ricerca per la genomica e la postgenomica, Fiorenzuola d'Arda), IBBR—Institute of Biosciences and Bioresources, Bari, MAC (Unità di ricerca per la maiscoltura, Bergamo), MPF (Unità di ricerca per il monitoraggio e la pianificazione forestale, Trento), OLI (Centro di ricerca per l'olivicoltura e l'industria olearia, Rende, Città S. Angelo e Spoleto), ORA (Unità di ricerca per l'orticoltura, Monsampolo del Tronto), ORL (Unità di ricerca per l'orticoltura, Montanaso Lombardo), ORT (Centro di ricerca per l'orticoltura, Pontecagnano), PLF (Unità di ricerca per le produzioni legnose fuori foresta, Casale Monferrato and Roma), QCE (Unità di ricerca per la valorizzazione qualitativa dei cereali, Roma), RIS (Unità di ricerca per la risicoltura, Vercelli), SCA (Unità di ricerca per i sistemi colturali degli ambienti caldo-aridi, Bari), SCV (Unità di ricerca per la selezione dei cereali e la valorizzazione delle varietà vegetali, S. Angelo Lodigiano), SEL (Centro di ricerca per la selvicoltura, Arezzo), SFM (Unità di ricerca per il recupero e la valorizzazione delle specie floricole mediterranee, Palermo), VIT (Centro di ricerca per la viticoltura, Conegliano), VIV (Unità di ricerca per il vivaismo e la gestione del verde ambientale ed ornamentale, Pescia). CNR Institutes belongs to CNR DISBA Department.

In 2004, with the approval of the FAO—International Treaty on Plant Genetic Resources for Food and Agriculture, one of the first binding global agreements on PGRFA in harmony with the CBD, the global agreement entered into force. It involves concrete obligations for the Contracting Parties regarding the conservation and documentation of species of agricultural interest, in order to facilitate access to them and share benefits arising from their use. For Italy, MiPAAF has the responsibility for the implementation of the FAO Treaty; MiPAAF entrusted the CREA-FRU with the scientific coordination of the actions for the collection, conservation, characterization, evaluation, and enhancement of PGR of agricultural interest, as defined in the specific project "Plant Genetic Resources/Implementation of the FAO Treaty", launched in 2004, that gives special priority to old and local varieties. The project involves 27 centers and Research Units belonging to the CREA, the former Institute of Plant Genetics of CNR in Bari (today IBBR), and, since 2008, 10 NGOs that have joined in the "Semi Rurali" Network

(Table 1). Starting in 2014, the CNR was involved at a high level through the Department of Biology, Agriculture, and Food Science, which holds many different plant and microbial collections through its network of institutions.

Sixty-five species are included in the project, of which 22 are listed in the Annex I of the FAO Treaty; the other 43 species are distinguished by their economic and strategic significance for Italy.

The Research Units of CREA store a large number of accessions (native and foreign material, old and new cultivars, populations, landraces, breeding lines, etc.), most of which are stored as seeds or in vivo; a small proportion of germplasm is also preserved through cryoconservation [13] and in vitro conservation.

The documentation of the characterization data regarding PGR is indispensable in making the results of the work available and to encourage the use of PGR in sustainable farming systems. The online catalogue "National Inventory of PGR stored ex situ in Italy", established in 2006 under the project managed by CRA-FRU is therefore proposed as a national platform to provide basic monitoring information (passport) as well as morphological and physiological data according to international standards. Currently, the database contains data on more than 30,000 accessions belonging to about 500 different species and stored in 44 Italian public institutions. The catalogue, thanks to its interactive nature, is constantly updated, a task accomplished independently by individual institutions, and therefore a constant increase in the number of accessions monitored and related information is expected.

1.4. PGR Stored by the Research Institutes of the CNR

The National Research Council (CNR) is a public research organization. The CNR scientific network consists of (a) Departments responsible for programming, coordination, and control; and (b) Institutes where the research activities are carried out.

The "Scienze Bio-Agroalimentari" Department (DISBA) consists of institutes that at various levels are involved in conservation and characterization of plant biodiversity and therefore hold collections of genetic resources. In particular, the Institute of Biosciences and Bioresources (IBBR) has, since 1970, a genebank, which was, at the time of its establishment, designed as the reference genebank for all the Mediterranean area. A large fruit tree collection is held by CNR-IVALSA and is also reported within the collections identified by CREA-FRU.

Currently, the DISBA has collections of animal genetic resources (pigs, cattle, sheep, but also insects and nematodes), model plants (*Arabidopsis*, *Medicago*, *Nicotiana*, etc.) and plants of food interest. In detail, the DISBA has the following collections: fruit trees (1860 accessions), *Citrus* (241 accessions), olive trees (about 2500 accessions), grapevine (119 accessions), forage plants (782 accessions of 83 species), vegetables, officinal plants and other species (1270 accessions of more than 200 species). The collections pertain to various institutes of CNR (IBBR, IVALSA, ISAFOM, IBAF, etc.). In particular, in the IBBR genebank in Bari, more than 65,000 accessions of over 600 different species are preserved. Most of the accessions belong to cereals and legumes, but also horticultural species and wild progenitors are maintained, including a living collection of artichoke. Of these accessions, more than 15,000 were directly collected by IBBR in collaboration with other national and international institutions (e.g., FAO, IBPGR (International Board for Plant Genetic Resources, today Bioversity International), etc.). These samples are also partially duplicated in other genebanks.

An initial investigation aimed at acquiring an overall picture of the situation was carried out by DISBA in 2008. However, there is not a common database which brings all the information together, yet. In addition, it is necessary to find a common and shared protocol for the conservation and utilization of the PGR stored.

1.5. Other Sources

Of course, in addition to the CNR and the CREA in Italy there are other institutions, both public (e.g., universities) and private (e.g., NGOs) that preserve plant germplasm collections of great value. The problem is that there is not yet a complete census of these institutions and of what precisely they preserve. Some initiatives have already arisen with a PGR census as the main aim. Among them there is a survey by the "Istituto Superiore per la Protezione e la Ricerca Ambientale"—ISPRA (Institute for Environmental Protection and Research)—which produced in 2010 a volume on the ex situ conservation of biodiversity of wild and cultivated plant species in Italy, including the state of the art, problems and actions to be taken [12].

In April 2013, the DISBA of CNR created BioGenRes, the Italian Network of Genetic Resources [14]. BioGenRes represents a first step towards the systematic rationalization and harmonization of national genetic resources, for the improvement of the agro-food industry and sustainable forest management. Finally, a project for the constitution of a national inventory is being conducted by CNR, CREA, INEA under the coordination of the Ministry of Agriculture, starting the National Inventory that is providing data to EURISCO. The European Search Catalogue for Plant Genetic Resources (EURISCO) provides information on 1.9 million accessions of crop plants and their wild relatives, preserved ex situ by almost 400 institutes. It is based on a network of National Inventories of 43 member countries and represents an important effort for the preservation of the world's agrobiological diversity by providing information about the large genetic diversity kept by the collaborating institutions. The central goal of EURISCO is to provide a one-stop-shop for information for the scientific community and for plant breeders. EURISCO contains both passport data and phenotypic data. EURISCO is being maintained on behalf of the Secretariat of the European Cooperative Programme for Plant Genetic Resources (ECPGR), in collaboration with and on behalf of the National Focal Points for the National Inventories.

2. A General Plan of Action for Italy to Improve the Safeguarding of Crop Genetic Resources

In the light of the above considerations, a plan of action for Italy should include the following main tasks:

1. To develop new (bio) informatics systems that can facilitate both the management of the utilization of stored genetic resources (e.g., finding duplicates of accessions, defining core collections), making them readily available, and doing work together on data of different nature (passport data, evaluation, images, GIS mapping, etc.) to help breeders to select the best parents for their breeding programs.

2. To develop (bio) informatics systems that will aid researchers to census the level of synonymy/duplication internal to the collections. Unwanted duplication may be due to obtaining the same genotypes from different sources, or from the fact that the same genotype is called by different names in different areas (a typical example is the olive germplasm).

3. To assess the level of safety duplication of the material stored, i.e., whether each sample has a "backup copy" stored at another center for the conservation and, if not, developing it also using innovative techniques of in vitro conservation.

4. To establish contacts and to formalize interactions with major international institutions for safeguarding plant biodiversity, such as, CGIAR (Consultative Group on Agricultural Research), Bioversity International, the European Network ECPGR (some important Italian genebanks only a few months ago joined to EURISCO and AEGIS, others have not yet!).

5. The main critical factor is the lack of a single national institution responsible for the conservation of all PGR of agricultural interest or of a coordinated germplasm system. This national institution should also have the task of coordinating activities by other organizations at national and regional level for the purposes of a correct policy of duplication of collected accessions. The accessions of many species of agricultural interest are disappearing quickly; traditional crops have almost been completely replaced by a few commercial varieties. The consequence is the decrease of genetic

variability in the fields. The survival of many genotypes is exclusively linked to their presence in collections. A lack of cooperation among the various institutions (public and/or private) involved in the conservation of PGR should be noted. It is also important to mention the lack of adequate and continuous funding for the care and maintenance of the collections, including characterization activities. Equally important are: (a) the difficulties in finding adequate space for new accessions (often indigenous material threatened by genetic erosion), especially for tree species; (b) lack of facilities for the proper arrangement of the material to be quarantined; (c) the great heterogeneity in the documentation of the accessions stored at the various institutions, with the consequent difficulty of harmonizing the data contained in the various databases maintained by individual institutions and often specific to only a few species of interest.

6.   Some additional actions to be taken for solving the most critical factors are: (a) to define and institutionalize a national institution for the conservation of PGR for agriculture; (b) to continue the work of collecting accessions in the national territory which are not yet included in public collections; (c) to continue and complete the morphological, agronomical, phytosanitary and molecular characterization of all stored accessions; (d) to improve, complete and harmonize the documentation of the stored material (e.g., census of facilities that operate the active conservation of PGR, census of species/varieties stored); (e) to define, for each crop, a *core collection*, in order to ensure the efficiency of evaluation and the conservation of essential genetic traits; (f) to carry out public awareness-raising activities for the safeguarding of PGR and to create awareness regarding the various potential uses of PGR and the importance of genetic variation within a given species; (g) to properly prepare the material, especially that under the FAO Treaty, for exchange with other institutions; (h) to create conditions for increasing the duration of the viability of accessions in seed storage (suitable climatic chambers for long-term storage); (i) to assess the conservation status of the material currently present in ex situ collections in order to effectively intervene on the endangered species from extinction; (j) to promote the use of the National Inventory as a general platform for documentation and access to data on the PGR stored ex situ in Italy. This will also facilitate the transfer of information into the various European (EURISCO and the European Central Crop Databases, ECCDBs) and global catalogues (WIEWS, Genesys).

3. Genetic Resources of the Main Vegetables Cultivated in Italy and Their Safeguarding

3.1. Italian Situation

Vegetable crops in Italy, covering a total area of about 530,000 hectares, belong to about 40 species, forming a very heterogeneous group. With the exception of tomato (123,000 hectares), potato (80,000 hectares), artichoke (49,000 hectares), fresh green bean, cauliflower, fennel, lettuce and melon (22,000–24,000 hectares each), the area of all other vegetables comprises only a few thousand hectares.

The conservation of genetic resources in the process of rapid and final extinction has become, for some decades, one of the most urgent objectives of genetic research applied to plants, including vegetables. In fact, the relentless progress of cultivation techniques can provide income gains only if they are applied to genotypes resistant to pests, suitably adapted to high fertilization, integral mechanization, chemical weed control, crop protection, and artificial substrates. Commercial distribution of vegetable seeds, which has almost completely replaced the seed harvested by the farmer himself, enhances improved cultivars and hybrids according to the requirements cited above, the presence of which in the market, as a result of the rapid varietal evolution, usually does not exceed three to four years. In addition, a new vegetable cultivar, to be profitable for the breeder, has to be protected: this is the reason why, beyond the undeniable merits, the F1 hybrids have become more widespread, and have drastically reduced the use of open-pollinated cultivars, the cost of multiplication of which is similar to that of hybrids, but their pay-back for the seed producer is much lower.

The seed industry is increasingly concentrated in the hands of a few multinational corporations; it engages mainly in obtaining F1 hybrids resulting from a narrow range of parental lines, or providing genetically improved crop varieties, consequently, to the preservation of only a small number of traditional cultivars of particular notoriety and gradually abandoning all the others. This has caused, and still causes a rapid loss of genetic variation. The old local populations (or landraces) perfectly adapted to their environment, the nowadays obsolete commercial cultivars, the lines already used in the work of breeding and today discarded, are, however, a wealth of unique genetic variability, the loss of which cannot be remedied. The collection, characterization, and conservation of genetic resources are, therefore, of particular importance, especially in the field of vegetable crops, of which Italy is historically very rich. To face the problems of genetic erosion, the "National Register of Horticultural Varieties" was established in Italy in the 1970s (Ministerial Decree of 17.07.1976) in which 726 local varieties called "ante '70" were recorded. Later, because of the constant negative feedback relating to the varietal identity of samples stored at seed industries responsible for their conservation and the lack of available subjects to carry out their maintenance in purity, it has come to a renewal of the above mentioned register that led to the cancellation of 326 varieties. To them should be added other 46 varieties cancelled due to lack of identity requirements and varietal homogeneity. Today, the new list includes both open-pollinated varieties (506 from the old list and 350 made after 1977), and 74 F1 hybrids from the old list and 490 hybrids registered after 1977. Seed companies or public institutions keep them in genetic purity. More details on the "National Register of Horticultural Varieties", its updated list and the implications for a variety being included in that list are here reported [6].

The promotion and development of local products is one of the most important agricultural policy strategies for the revitalization of the Italian agricultural economy, in particular for the South, where agriculture often does not have the technical and economic conditions necessary to compete with the more advanced agricultural systems or to cope with the competition from foreign countries producing at lower costs. The promotion of local products also contributes to the preservation of agrobiodiversity: a large amount of crop germplasm would be lost (or would have already disappeared) if not properly valued and promoted through collective marks (PDO, PGI, AS or Attestation of Specificity, GTS or Guaranteed Traditional Specialty), which represent important regulatory instruments to protect consumers and to support small and medium farms.

The whole Italian territory, but particularly inland areas of southern Italy where small family-owned farms still exist, is particularly rich in vegetable germplasm represented by different landraces clearly distinguishable from other similar cultivars (for morphological characteristics, sensorial, etc.) and closely linked to the historical memory of their places of origin [5].

3.2. Main Safeguarding Problems

The numerous scientific activities undertaken so disconnected from the actors in the territory, threaten to undermine the work already carried out with considerable financial resources at regional, national, and EU level. Therefore, it is necessary that all steps of recovery, characterization, conservation, and exploitation are taken only and exclusively in agreement or at the suggestion of local actors, public or private, located and operating in the territory concerned. In particular, a lack of homogeneity of methodological approaches adopted in the collection, classification, measurement, and characterization of the material can be observed. In addition, the exploration of the territory is not always followed by adequate preservation of the collected material.

The lack of coordination has often led to overlapping of initiatives and a confusion of roles which would be appropriate to bring order, to better leverage the work already conducted, and to efficiently address future activities. In addition, the lack of appropriate funding necessary to develop further the activities of ex situ conservation, with costs generally high, has brought more problems and confusion in the work.

The evaluation activities of the stored material and studies on the genotype × environment interaction on the most interesting landraces are insufficient. In the same way the knowledge about the most effective methods of ex situ conservation is incomplete (e.g., for some crops there are no experimental data on the best conservation parameters and conditions). The currently existing genebanks have played and continue to play, an important role in the collection and preservation of plant genetic resources, but it is equally true that ex situ conservation alone does not guarantee the actual conservation of the resources and their durable use. Another important priority is to define the risk threshold beyond which the varieties are considered at risk of extinction and therefore would need protection. These thresholds must be recognized and shared by all scientific and non-scientific subjects working in this field.

3.3. Some Actions to Take for Solving the Most Critical Factors of Vegetable Landraces

It is very important to guarantee the maintenance and management of existing collections and to survey and to conduct a complete census of ecotypes of vegetable species originating and/or historically present in the regional agricultural areas, in the way of what regional laws foresee. It is of great practical importance to collect morphological, chemical, agronomic, and molecular data for the widest possible characterization of germplasm, in order to identify the potentially most interesting traits, such as the production of bioactive compounds (e.g., vitamins, fiber, minerals, antioxidants, enzymes, etc.) important in the prevention of many diseases.

It might be useful to evaluate existing genebanks, in terms of functionality and capabilities, and to study and develop 'specific' methodologies and equipment for the seed preservation, to ensure the integrity of the genetic material in the long term. For example: (a) are we sure that we can store in the same storage room about 1000 different species (this is common in many genebanks) thinking that the climatic parameters used, are the best for all of them? (b) many wild relatives of pulses need specific rhizobium to grow but usually genebanks do not care for this; (c) the same for plants that need specific pollinators, no part of genebank is dedicated to them; (d) almost no genebanks monitor the genetic erosion which occurs in them!

Some multiplication problems can be solved by improving the study on micropropagation techniques, which for many vegetables could be a great help, as they require less space and costs to store and to periodically rejuvenate the material.

An effective and unique 'official' database of genetic material collected, possibly on-line, is essential together with evaluation of the agronomic and commercial potential of the best landraces.

To perform better actions for a targeted breeding we have to improve the quality and usability of information about evaluation data regarding accessions in the collections of germplasm. We must increase the spread of the technological and scientific results obtained during the investigation on the best characteristics of traditional products under investigation, and pilot actions to diffuse the cultivation of the most typically neglected vegetable landraces.

Additional useful actions might be: (a) draft cultivation specifications and application for release of protection collective marks; (b) trade promotion activities of neglected local vegetables through awareness and information campaigns; (c) implement the collections through exchange with other research institutes and Italian and foreign genebanks, seeking to create synergies and ways of interaction as part of the multiplication and rejuvenation of the seed, in order to optimize the ex situ conservation of germplasm (this is the aim of the AEGIS, the European Genebank Integrated System, of which Italy is a member); (d) to prepare guidelines for the definition of a program of activities for the protection of national biodiversity, to be carried out according to the indicators for the quantification of the specific objectives of the Rural Development Programme 2014–2020 (in this respect the national guidelines of the "Piano Nazionale sulla Biodiversità di Interesse agricolo", is too general); (e) to define management protocols nationwide standardized for the ex situ conservation of the main local varieties; (f) to create networks of "guardian farmers", such as contacts and responsible for the renewal and multiplication of biodiversity products recovered in the territory, recognizing the work so "loving" which they have done over the years, as defined also by the National Law on Agriculture and Food Biodiversity N.194, of 2015; (g) to ensure the economic sustainability of conservation actions (guardian farmers or any person involved in safeguarding of germplasm); (h) to stimulate multifunctionality of farms as a tool for possible economic sustainability of conservation actions (e.g., farmhouses offering product of landraces produced on-site).

4. Conclusions

This study attempted, above all, to provide a summary, with a strictly scientific basis, on the ex situ conservation of Italian agricultural biodiversity and, therefore, could represent a small contribution to the national strategy for the protection of its agrobiodiversity in general.

In addition to the technical and methodological problems, however, the ex situ conservation is also affected by a general unavailability or shortage of funds, which limits its development. This phenomenon applies to the majority of genebanks around the world and is accompanied in many cases by a lack of interest of policy makers in the subject. There is no doubt that the focus on ex situ conservation, very strong in the 1960s and 1970s, when the first genebanks arose, has been gradually reduced.

According to FAO (2010), the world's genebanks store ex situ ca. 7.4 million accessions of cultivated species (e.g., cereals, legumes, vegetables, fodder, officinal, medicinal, aromatic, etc.), wild relatives of the cultivated species, and other wild species, threatened by genetic erosion and/or extinction.

In the future there will be an increasing need to develop sustainable agricultural systems, for both food and energy and to preserve cultivated and wild species against genetic erosion. The genebanks can definitely play a decisive role, complementary to the in situ conservation (incl. on-farm conservation) and to more careful territory planning. In this perspective, a greater economic effort is desirable aimed at the development of research, the maintenance of genebanks, and the continuous monitoring of the state of the collections. A political and normative commitment in this sense is crucial, supporting the ex situ conservation (at the moment this support is still ineffective). In general, a greater involvement by governments of different countries is desirable to support the networks of genebanks and the activation of participatory systems that involve the entire chain of production, from farmers to end users, in order to develop territory management seriously and concretely oriented to sustainability.

5. Case Study: "Mugnolicchio": A Neglected Race of *Brassica Oleracea* L. from Altamura (Italy)

The *Brassicaceae* plants are among the most consumed vegetables in the world. They feature a large biodiversity, in which landraces and primitive cultivars still play a major role in the cultivation systems of many countries. Many cultivated *Brassicaceae* and especially broccoli are rich in antioxidant compounds that play a key role for human health especially in traditional cuisine [15]. Italy is widely regarded as the center of genetic diversity for several cultivated *Brassicaceae*, such as *B. oleracea* L. var. *botrytis* L. (cauliflower) and var. *italica* Plenck (broccoli). Therefore, many specific exploration missions have been carried out in Italy to collect *Brassica* germplasm both cultivated [16,17] and wild [18,19].

This rare landrace of *Brassica oleracea* was found in the 2014 (Figures 1–3) in Altamura (Bari) and for the first time a preliminary characterization was made. It is called "Mugnolicchio" or "Mignolicchio" and is cultivated traditionally in the Altamura area (Apulia region, southern Italy).

Figure 1. Different big inflorescences of "Mugnolicchio" in the same plant.

Figure 2. Plant and inflorescences of "Mugnolicchio".

Figure 3. Flowers of "Mugnolicchio".

"Mugnolicchio" is similar to the broccoli of which, according to recent investigations, it is (probably with "Mugnoli" of Salento—Figure 4), the progenitor from which the latter were selected.

However, only a specific genetic study, considering all together its wild and cultivated relatives, will clarify if "Mugnoli" is an ancestor or whether it is a parallel development [20].

Figure 4. "Mugnoli" of Salento.

Morphologically it is clearly distinguishable from the broccoli (Figures 5–7) for the smaller and less compact inflorescence; the single flowers of the "Mugnolicchio" are white, larger and with bracts larger than those of broccoli. Its organoleptic characteristics are peculiar too and often people prefer it to broccoli. There are many traditional recipes with "Mugnolicchio" in the AltaMurgia area, all aimed to extoll its sweet and aromatic flavor.

"Mugnolicchio" is a relict landrace because in the area of Altamura (Ba) its cultivation is decreasing (Figures 8 and 9). The standardization of modern cultivars caused a rapid decline of this landrace unable to compete in the market because of its small inflorescences and lack of scalar production. Nowadays it is still produced by small farmers for family use, and very much appreciated by local people.

Figure 5. Commercial broccoli variety.

Figure 6. Height difference from "Mugnolicchio"(left) and broccoli (right).

Figure 7. Cultivation of "Mugnolicchio".

Figure 8. Cultivation of "Mugnolicchio".

It is still cultivated in small plots of land by some horticulturists. It is sown in August and transplanted in autumn in order to collect the inflorescences from March onwards. The plant can be grown for four years, after that it is replaced. In the past, farmers sowed this crop to separate their own plots from neighboring areas, as a kind of demarcation. Some plants were also sown in April for the exclusive use of the fleshiness of the leaves in summer, cooking them with pasta, mainly when in the middle of summer there are no other cultivated *Brassicacae*. There are two morphological types cultivated in the same area. Until now only the morphotype with smooth and slightly lobed leaves is stored in a genebank (i.e., in Bari at IBBR-CNR). One morphotype is characterized by smooth and slightly lobed leaves (Figure 9), the other one by fleshy and very lobed leaves (Figure 10). This last morphotype is probably the typical landrace of the past, because the characteristics of the leaves would make it more appreciated for food.

Figure 9. Morphotype with smooth and slightly lobed leaves.

Figure 10. Morphotype with fleshy and very lobed leaves.

The case of "Mugnolicchio" is only one out of a number of other examples of old Italian landraces that are now being broadly cultivated, as the lentil "Lenticchia di Altamura" [21], the common beans "Fagioli di Sarconi" [22] and "Fagiolo poverello bianco di Rotonda" [23], the scarlet eggplant (*Solanum aethiopicum* L.) "Melanzana rossa del Pollino" [24], the eggplant "Melanzana Bianca di Senise" [25], the old agroecotypes of potato of the Pollino National Park [26], hulled wheat (*Triticum dicoccon* Schrank and *T. spelta* L.) [26,27], etc.

Author Contributions: K.H. and G.L. conceived and designed the study; all authors performed the exploration and collecting missions used as source of data and information to write the article; K.H. analyzed the data; G.L., K.H., V.M., and P.D. wrote the paper.

Acknowledgments: This study was supported by the project "Implementazione del Trattato Internazionale FAO sulle Risorse Genetiche Vegetali per l'Alimentazione e l'Agricoltura' (RGV-FAO) funded by the Italian Ministry of Agriculture (MiPAAF).

Conflicts of Interest: The authors declare no conflict of interest.

References

1. Hammer, K.; Knüpffer, H.; Laghetti, G.; Perrino, P. *Seeds from the Past. A Catalogue of Crop Germplasm in South Italy and Sicily;* Germplasm Institute of C.N.R.: Bari, Italy, 1992; pp. II–173.

2. Hammer, K.; Knüpffer, H.; Laghetti, G.; Perrino, P. *Seeds from the Past. A Catalogue of Crop Germplasm in Central and North Italy;* Germplasm Institute of C.N.R.: Bari, Italy, 1999; pp. IV–254. ISBN 88-900347-0-X.

3. Hammer, K.; Laghetti, G. *Small Agricultural Islands and Plant Genetic Resources—Le Piccole Isole Rurali Italiane;* IGV-CNR: Bari, Italy, 2006; pp. X–246. ISBN 88-900347-4-2.

4. Hammer, K.; Laghetti, G.; Pignone, D. *Linguistic Islands and Plant Genetic Resources—The Case of the Arbëreshë;* ARACNE: Rome, Italy, 2011; ISBN 978-88-548-3958-8.

5. Montesano, V.; Negro, D.; Sarli, G.; Logozzo, G.; Spagnoletti Zeuli, P. Landraces in inland areas of the Basilicata Region Italy: Monitoring and perspectives for on farm conservation. *Genet. Res. Crop. Evol.* **2012**, *59*, 701–716. [CrossRef]

6. Ministero delle Politiche Agricole Alimentari e Forestali. *Linee Guida per la Conservazione e la Caratterizzazione della Biodiversità Vegetale, Animale e Microbica di Interesse per L'agricoltura;* INEA—Piano Nazionale sulla Biodiversità di Interesse Agricolo: Roma, Italy, 2013.

7. Hammer, K.; Laghetti, G. Genetic erosion—Examples from Italy. *Gene. Res. Crop Evol.* **2005**, *52*, 629–634. [CrossRef]

8. CREA. Available online: http://www.crea.gov.it/?lang=en (accessed on 30 June 2018).

9. CNR. Available online: https://www.cnr.it/it (accessed on 30 June 2018).

10. Aitken-Christie, J.; Kozai, T.; Smith, M.A.L. (Eds.) Glossary. In *Automation and Environmental Control in Plant Tissue Culture;* Kluwer Academic Publishers: Dordrecht, The Netherlands, 1995; pp. IX–XII.

11. Germanà, M.A.; Hafiz, I.A.; Micheli, M.; Standardi, A. Preliminary research on conversion of encapsulated somatic embryos of *Citrus reticulata* Blanco, cv. Mandarino Tardivo di Ciaculli. *Plant Cell Tissue Organ. Culture* **2007**, *88*, 117–120.

12. Piotto, B.; Giacanelli, V.; Ercole, S. (Eds.) La conservazione ex situ della Biodiversità delle Specie Vegetali Spontanee e Coltivate in Italia: Stato dell'Arte, Criticità e Azioni da Compiere. Manuali e linee guida ISPRA: Roma, Italy, 2010; ISBN 978-88-448-0416-9.

13. Lambardi, M.; de Carlo, A. Tecniche ed applicazioni della criogenia alla conservazione ed al risanamento di germoplasma vegetale. *Italus Hortus* **2009**, *16*, 79–98.

14. BioGenRes. Available online: http://www.biogenres.cnr.it (accessed on 30 June 2018).

15. Nicoletto, C.; Santagata, S.; Pino, S.; Sambo, P. Antioxidant characterization of different Italian broccoli landraces. *Horticult. Bras.* **2016**, *34*, 74–79. [CrossRef]

16. Laghetti, G.; Hammer, K.; Olita, G.; Perrino, P. Collecting vegetable crops in Basilicata, Italy. *Plant Genet. Resour. Newslett.* **1993**, *96*, 35–37.

17. Massie, I. *Report on Research Trip to Italy to Assess Genetic and Ecogeographic Variation and Genetic Erosion of Cauliflower and Broccoli Crops and to Collect Seed Samples;* Horticulture Research International, Wellesbourne: Warwick, UK, 1993.

18. Gustafsson, M.; Gomez-Campo, C.; Perrino, P. Germplasm conservation of the wild Mediterranean *Brassica* species. *J. Swed. Seed Assoc.* **1986**, *26*, 347–360.

19. Perrino, P.; Pignone, D.; Hammer, K. The occurrence of a wild *Brassica* of the *oleracea* group (2n = 18) in Calabria (Italy). *Euphytica* **1992**, *59*, 99–101. [CrossRef]

20. Laghetti, G.; Martignano, F.; Falco, V.; Cifarelli, S.; Gladis, T.; Hammer, K. "Mugnoli": A neglected race of *Brassica oleracea* L. from Salento (Italy). *Genet. Resour. Crop Evol.* **2005**, *52*, 635–639. [CrossRef]

21. Laghetti, G.; Volpe, N.; Sonnante, G.; Pignone, D.; Sonnante, G. On the old Apulia lentil agroecotype 'Lenticchia di Altamura' (Apulia, Italy). *Italus Hortus* **2006**, *13*, 467–471.
22. Piergiovanni, A.R.; Brandi, M.; Cerbino, D.; Olita, G.; Laghetti, G. The agro-ecotypes of common bean from Sarconi and Rotonda (PZ). Results of a triennal agronomic and biochemical study. (Gli agro-ecotipi di fagiolo di Sarconi e Rotonda (PZ). Risultati di una sperimentazione agronomica e biochimica triennale.). In Proceedings of the V National Congress on Biodiversity "Biodiversità e Sistemi Ecocompatibili", San Leucio (CE), Italy, 9–10 September 1999; Volume 13, pp. 571–578.
23. Brandi, M.; Cerbino, D.; Laghetti, G.; Piergiovanni, A.R.; Olita, G.; Rizzi, R.; Martelli, S. Una carta di identità per il fagiolo di Sarconi e Rotonda. *Inf. Agrar.* **1998**, *27*, 55–61.
24. Laghetti, G.; Hammer, K.; Brandi, M.; Cerbino, D.; Olita, G.; Perrino, P. Ritrovamento di una coltivazione di melanzana africana. *Inf. Agrar.* **1995**, *39*, 52.
25. Cerbino, D.; Laghetti, G.; Piergiovanni, A.R. Studi morfologici e biochimici per rilanciare la 'Melanzana Bianca di Senise'. *Agrifoglio* **2010**, *37*, 38–39.
26. Cerbino, D.; Illiano, M.; Cirigliano, M.; Di Napoli, A.; Zienna, P.; Gallo, S.; Laghetti, G.; De Lisi, A.; Direnzo, P.; Giunta, R.; et al. *Le Antiche Varietà di Patata del Pollino*; Laghetti, G., Cerbino, D., Eds.; I Quaderni dell'ALSIA n.12/2016—Supplement to N. 55 of Agrifoglio; ALSIA: Matera, Italy, 2016; Volume 124, ISBN 978-88-95110-20-2.
27. Perrino, P.; Volpe, N.; Laghetti, G. "Lucanica" e "Forenza": Due nuove varietà di farro. *Inf. Agrar.* **1996**, *47*, 34–35.

agriculture

MDPI

Article

Issues and Prospects for the Sustainable Use and Conservation of Cultivated Vegetable Diversity for More Nutrition-Sensitive Agriculture

Gennifer Meldrum [1,*], Stefano Padulosi [1], Gaia Lochetti [1], Rose Robitaille [1] and Stefano Diulgheroff [2]

[1]   Healthy Diets from Sustainable Production Systems Initiative, Bioversity International, Via dei Tre Denari, 472/a 00054 Maccarese, Italy; s.padulosi@cgiar.org (S.P.); g.lochetti@cgiar.org (G.L.); r.robitaille@cgiar.org (R.R.)
[2]   Plant Production and Protection Division, Food and Agriculture Organization of the United Nations, Viale delle Terme di Caracalla, 00153 Rome, Italy; Stefano.Diulgheroff@fao.org
*   Correspondence: g.meldrum@cgiar.org; Tel.: +39-06-6118272

Received: 1 June 2018; Accepted: 6 July 2018; Published: 9 July 2018

Abstract: Traditional vegetables are key assets for supporting more nutrition-sensitive agriculture under climate change as many have lower water requirements, adaptation to poor quality soils, higher resistance to pests and diseases, and higher nutritional values as compared to global vegetables. The effective use of traditional vegetables can be challenged however by lack of information and poor conservation status. This study reviewed the uses, growth forms and geographic origins of cultivated vegetables worldwide and the levels of research, ex situ conservation, and documentation they have received in order to identify gaps and priorities for supporting more effective use of global vegetable diversity. A total of 1097 vegetables were identified in a review of the Mansfeld Encyclopedia of Agricultural and Horticultural Plants, including species used for leaves ($n = 495$), multiple vegetative parts ($n = 227$), roots ($n = 204$), fruits or seeds ($n = 90$), and other parts like flowers, inflorescences, and stems ($n = 81$). Root vegetables have received significantly less research attention than other types of vegetable. Therophytes (annuals) have received significantly more attention from research and conservation efforts than vegetables with other growth forms, while vegetables originating in Africa ($n = 406$) and the Asian-Pacific region ($n = 165$) are notably neglected. Documentation for most vegetable species is poor and the conservation of many vegetables is largely realized on farm through continued use. Supportive policies are needed to advance research, conservation, and documentation of neglected vegetable species to protect and further their role in nutrition-sensitive agriculture.

Keywords: traditional crops; cultivated vegetables; neglected and underutilized species; nutrition; climate change adaptation

---

1. Introduction

Vegetables are important sources of micronutrients, including vitamins, minerals, antioxidants and fibre needed to conduct a healthy and productive life [1,2]. They are among the most diverse, colourful and tastiest foods, and are strategic for reaching balanced diets and reducing the incidence of severe health ailments [1,3]. Current nutrition guidelines recommend consumption of at least 400 g (5 portions) of fruits and vegetables per day [4], yet a large proportion of individuals do not meet these requirements, which is contributing to rates of malnutrition and the rise of non-communicable diseases around the world [5–7].

Links between agriculture and nutrition are well documented and historic trends in agricultural development are acknowledged to have contributed to current diet insufficiencies [8,9].

The calorie-centric approach focused on enhancing yields of a few staple cereals through the Green Revolution has led to a profound loss of diversity in agriculture and food systems. Rice, wheat and maize account for 51 percent of plant-based caloric intake and 42 percent of the total food supply (kcal/capita/day) in human diets, meanwhile their cultivation covers 40% of arable land globally [10]. In comparison to the major cereals, investments in research and development for vegetables have been negligible and focused mainly on a small basket of globally-important crops [11]. Production of vegetables and fruits is currently insufficient to meet the needs of the human population, with supply deficits of 22% on average and up to 58% in low-income countries [12]. Value chains for vegetables are, moreover, poorly developed in many places, which limits their accessibility to consumers [13–16]. Enabling access to vegetables at an affordable price is both an emerging priority and a challenge for policy makers as populations become increasingly urbanized and reliant on purchased foods [17–20].

In recognizing the need for nutrition-sensitive agriculture and food systems, greater vegetable production and use are being called for and promoted through horticulture, home gardens, urban and peri-urban agriculture, agroforestry, and school feeding programmes, among other approaches [21–28]. As well as supplying nutritious food, vegetable production is also recognized as a profitable sector that can support income generation [11,29]. While having great potentials, vegetable cultivation also faces important agronomic challenges and limitations, especially with regards to water availability, soil fertility, and pest and disease control. It is highly sensitive to climate change [30–33] and can enhance vulnerability of producers in water limited areas [34,35], as well as exposure to harmful chemical inputs [34,36–38]. To ensure a holistically sustainable development trajectory, the transformation towards vegetable-rich production and food systems should also support climate change adaptation and protection of human and environmental health [39].

Traditional vegetables are an important asset for meeting this challenge as many have high nutritional value, low water requirements, adaptation to poor quality soils, and good resistance to pests and diseases [40–45]. Several indigenous leafy vegetables of Africa present an optimal source of nutrients such as β-carotene, folate, iron, calcium, zinc, proteins and dietary fibre [46–49], while showing lower water use and higher water use efficiency compared to introduced vegetables such Swiss chard (*Beta vulgaris* subsp. *Vulgaris* L.) [50]. Chaya (*Cnidosculus aconitifolius* (Mill.) I.M. Johnst) is a shrub native to Mesoamerica that thrives with few inputs in arid conditions and produces leaves with two to three times the nutrient value of spinach and lettuce [51,52]. The greater content of important macro and micronutrients found in many traditional vegetables is partly the result of crop improvement favouring selection of traits such as high yield, shelf life, and appearance, while neglecting traits such as vitamin and mineral content [44,53,54]. Traditional vegetables can also require relatively fewer labour and economic inputs compared to global vegetables, meaning they present lower risks of financial losses for small farmers [55]. Because of their nutritional values and local adaptation, there is a growing body of literature highlighting how greater production and consumption of traditional and indigenous vegetables can support nutrition security and incomes [46,56–59]. However, more research is needed to clarify and leverage the roles and potentials of specific species, as the complex relationships between nutritional yields, water availability, and soil quality [60,61] remain underexplored for many species, as do their acceptability to consumers and capacities for integration in value chains.

Lack of knowledge and research generally challenges the promotion and use of traditional vegetables. Similar to other neglected and underutilized crop species, traditional vegetables are characterized by limited research efforts, breeding efforts, germplasm characterization, knowledge on species distribution and production levels, and representation in ex situ collections [62]. A dearth of information and poor awareness may allow useful species to be overlooked through a vicious cycle of neglect and underutilization. Declining use and eroding knowledge of traditional vegetables has been observed in many places around the world, which threatens their persistence into the future and limits the delivery of their benefits to society [63,64]. One million accessions are kept in world gene banks for vegetable crops but they mainly cover a small number of commodity crops (viz. tomatoes, capsicums, melons and cantaloupe, brassicas, cucurbits, alliums, okra, and eggplant) and crops with important non-vegetable uses such as grain, pulse or fibre [65]. The state of conservation of traditional vegetables remains largely underexplored and poorly documented but many are likely conserved primarily through continued cultivation on-farm, which is a fragile situation where their use is declining [45].

This study aimed to shed light on the diversity of cultivated vegetables worldwide and to highlight opportunities to leverage neglected and underutilized species for more nutrition-sensitive agriculture. A database of cultivated vegetable species was compiled by review of the Mansfeld Encyclopedia of Agricultural and Horticultural Crops [66] and trends and gaps for their research, conservation, and documentation were evaluated in relation to their uses, growth forms, and geographic origins. The results reveal priority areas for research and development which can help to build the knowledge base and strengthen the conservation of vegetable diversity to support its integration in more nutrition-sensitive production systems.

2. Materials and Methods

A database of cultivated vegetable species was compiled for the study based on the 3rd edition of the Mansfeld's Encyclopedia of Agricultural and Horticultural Crops [66]. This resource covers more than 6040 species cultivated by humans, excluding ornamentals. The list of species has been *compiled through comprehensive reviews of the scientific literature and contributions from botanical institutes, gardens, and research centers around the world* [66,67]. It is among the most thorough databases of cultivated plants at the global level and has been used in previous assessments of cultivated plant diversity [68,69]. Existing global reviews of vegetable species (e.g., [70,71]) do not explicitly include all minor and traditional vegetables cultivated in local food systems. The Mansfeld Encyclopedia was selected because it best matched the objective and scope of the study to consider all cultivated vegetables, including minor species, while following a consistent format suitable for global level analysis.

Any plant part consumed for food that is not a mature fruit or seed is by definition a vegetable, meanwhile fruits (and legume pods) prepared in salads and savoury dishes are also considered vegetables in a culinary sense [72]. All plant species with vegetative parts consumed or for which fruits and seeds were explicitly mentioned to be consumed as a vegetable in the Mansfeld Encyclopedia were included in the database (Database S1). Non-vegetable uses of the seed (as cereals, grains or pulses) and fruit (consumed as a sweet or tart snack, dessert or side dish) were noted. The distinction between vegetable and non-vegetable uses for the fruits and seeds was challenging and arbitrary in some cases because it is based on perception and preparation. For this reason, we acknowledge that some inconsistencies have likely occurred in the database. To enable comparisons with other databases, the list of cultivated vegetables was standardized to accepted synonyms on the Plant List (http://www.theplantlist.org), which is a noble attempt toward a comprehensive online database of all plant species initiated by Royal Botanic Gardens, Kew, and the Missouri Botanical Garden [73]. The process of standardizing the synonyms was automated using the Taxonstand package in R [74]. Unresolved species were maintained that did not have other potential synonyms in the list.

2.1. Species Characterization

For all the species in the database, the specific part/s used as a vegetable were scored. In cases when the parts utilized as vegetable were unclear or unspecified in the Mansfeld Encyclopedia, additional credible data sources were consulted for clarification. Five distinct groups of vegetables were defined based on their use: "leafy vegetables" that are used for their leaves and which may also be used for their shoots; "root vegetables" for which roots, tubers, rhizomes, corms or bulbs are used; "fruit and seed vegetables" for which the fruit, pods, or fresh seeds are used as vegetables; "other vegetables" used for other specific parts such as flowers, stems, and shoots, and "multiuse vegetables" which, in contrast to the previous groups, have multiple parts used as vegetables. This grouping was made with reference to exploratory analyses with multiple correspondence analysis (MCA) and hierarchical clustering in the FactoMineR package in R.

In addition to plant uses, the geographic region of origin for each vegetable species was also documented. Each species was classified based on the notes in the Mansfeld Encyclopedia regarding countries and ranges of cultivation into the twelve cradles of agriculture and centres of diversity proposed by Zeven and Zhukovsky [75]. Following a similar process as for classifying species into use typologies, species were assigned into five groups reflecting common geographic origin. Species were grouped together that had a clear origin in (1) the Americas; (2) Saharan and sub-Saharan Africa; (3) the region spanning Europe, the Mediterranean, the Near East, and Central Asia; and (4) Asia, Australia and the Pacific Islands. A fifth group included wider ranging species whose origin was unclear or which spanned several regions.

Literature searches were furthermore performed for each vegetable species to classify them by growth form. The Raunkiaer life form system [76] as modified by Govaerts and colleagues [77] was applied, which is a fairly simple and widely used classification for plants that relates to many aspects of plant ecology, including reproductive mode, lifespan, and associated climate [78,79]. The Raunkiaer life forms are defined based on how species survive in unfavourable seasons and particularly how well the vegetative buds are protected. Further detail on the classification is provided in Table 1. In addition to classifying the cultivated vegetables by these 10 life-forms, additional characterization as climbing, succulent and parasitic plants was followed as per Govaerts et al. [77].

Table 1. Growth form classification of plant life forms sensu Raunkiaer [76] and Govaerts et al. [77].

Life Form	Characteristics
Phanerophytes	Persistent woody stems and buds that project 3 m or more above the soil. Includes trees and large shrubs, e.g., *Moringa oleifera* Lam.
Nanophanerophytes	Woody, persistent stems, with buds located between 0.5 m and 3 m above ground level. Includes smaller shrubs, e.g., *Cordyline fruticosa* (L.) A. Chev.
Herbaceous phanerophytes	Herbaceous stems projecting more than 0.5 m above ground level that persist for several years. Includes many tropical species, e.g., *Musa acuminata* Colla.
Chamaephytes	Persistent stems that are herbaceous or woody with buds located above soil level, but never by more than 0.5 m. Includes dwarf shrubs and some perennial herbs, e.g., *Aloe macrocarpa* Tod.
Hemicryptophytes	Herbaceous stems that often die-back during unfavourable seasons with surviving buds placed on (or just below) soil level. Includes many biennial and perennial herbs, including those in which buds grow from a basal rosette, e.g., *Lactuca sativa* L.
Geophytes	Stems that die back during unfavourable seasons with the plant surviving as a bulb, rhizome, tuber or root bud, e.g., *Daucus carota* L.
Therophytes	Complete their entire life-cycle during the favourable season and survive the unfavourable season as a seed. This group includes all annual herbs, e.g., *Corchorus olitorius* L.
Epiphytes	Growing buds occur on another plant, e.g., *Peperomia pereskiifolia* (Jacq.) Kunth.
Helophytes	Surviving buds are buried in water-saturated soil, or below water-level, but with flowers and leaves that are fully emergent during the growing season. Includes many marsh plants and emergent aquatic herbs, e.g., *Typha latifolia* L.
Hydrophytes	Fully aquatic herbs in which surviving buds are submerged, or buried in soil beneath water. Stems and vegetative shoots grow entirely underwater and leaves can be submerged or floating, but only the flower-bearing parts may be emergent, e.g., *Vallisneria natans* (Lour.) H.Hara.

2.2. Indicators of Neglect

Three indicators of neglect from research and development were assessed for each of the cultivated vegetable species following a similar approach to Galluzzi and Lopez Noriega [62], as described below.

Firstly, the number of records in Google scholar was used as an indication of research effort devoted to the species. Google scholar is a well utilized and robust index of academic literature from multiple disciplines that concern agriculture, including the social sciences and life sciences [80]. Google Scholar has some disadvantages compared to more controlled databases, including full-text rather than field-level search, lack of controlled vocabulary [81] and duplicated-records [82]. However, the coverage and accuracy has greatly improved over time and the index has some important advantages. In particular, accessibility was an important criterion for this review and a primary reason why Google Scholar was preferred over subscription based databases [83]. Comparisons of Google Scholar search results to other databases show a strong overlap [80]. A search was conducted for each species including the genus and species epithet as required words to be included along with at least one of the words "nutrition", "food" or "vegetable". This specification was made to help limit the search results to food uses and exclude studies relating mainly to pharmacology and other aspects. The search was limited to the previous 20 years (1997–2017) and was performed for the established synonym(s).

Secondly, the number of accessions maintained in ex situ germplasm collections worldwide was assessed using the World Information and Early Warning System on Plant Genetic Resources for Food and Agriculture database (WIEWS). The WIEWS database provides access to official figures on the number of plant genetic resources for food and agriculture secured in either medium- or long-term conservation facilities, as part of the monitoring of the implementation of the Second Global Plan of Action for Plant Genetic Resources for Food and Agriculture, and the plant component of Sustainable Development Goal indicator 2.5.1. The total number of accessions maintained for each cultivated vegetable species was queried by searching the established synonym and additional common synonyms.

Thirdly, production data from the Food and Agriculture Organization of the United Nations Statistical Databases (FAOSTAT) were used as an indicator of knowledge on species distribution and production levels. These agricultural statistics are reported by member nations and collected from agricultural yearbooks and other publications [84]. The data are not always based on direct observations, which results in some inconsistencies. Nonetheless, they a rare standardized source of cropping information and a pillar for global analyses of crop production [84,85]. Data for vegetables in FAOSTAT primarily concern those grown for human consumption in field and market gardens, while excluding those grown in small family gardens for household consumption. It is noted that significant gaps in the coverage of FAOSTAT would naturally exclude some of the vegetable species covered in this review. However, as this database represents a standard for agricultural production statistics and reflects on the detail of data collected by nation states, it was considered as a reasonable indicator of documentation (and knowledge) of species distribution and production levels. The number of countries with official data for different vegetables over the past 20 years (1997–2016) was assessed.

2.3. Relating Indicators of Neglect to Use, Growth Form and Region of Origin

The relationship between the three indicators of neglect (number of Google scholar records, number of accessions, and documentation in FAOSTAT) and species characteristics (region of origin, growth form, and vegetable and non-vegetable uses) were explored using statistical analyses. Welch's Analysis of Variance (ANOVA) was applied to test how the number of Google scholar records and accessions maintained in world gene banks relate to species characteristics. Welch's ANOVA is suitable for cases with unequal variance and sample sizes between groups but it assumes the sample conforms to a normal distribution, which was achieved by log-transformation. Following a similar approach to other researchers [86,87], post hoc pairwise comparisons were made using Games and Howell tests, which have similar assumptions and are consistent with Welch's ANOVA [88,89]. Chi-square tests were similarly used to assess how the probability of being included in FAOSTAT (either as a specific species or as part of a group of vegetables) related to species characteristics. In this case, Fisher exact tests were applied for post hoc pairwise comparisons. All analyses were performed using R version

3.4.3 (R Foundation for Statistical Computing, Vienna, Austria) in R Studio version 1.1.383 (RStudio, Inc., Boston, MA, USA).

3. Results

A total of 1097 cultivated vegetable species from 133 families and 544 genera were identified in the study. The families with greatest number of cultivated vegetable species were the Leguminosae ($n = 127$), Compositae ($n = 85$), Dioscoreaceae ($n = 56$), Amaranthaceae ($n = 45$) and Araceae ($n = 44$). The genera with the most cultivated vegetable species were the Dioscorea ($n = 54$), Solanum ($n = 26$) and Allium ($n = 26$). Almost all species had accepted synonyms but 32 species names were unresolved.

3.1. Uses

Various plant parts are used as vegetables including above-ground vegetative structures like leaves (58%), shoots (15%), and stems (3%), underground vegetative structures such as tubers (12%), rhizomes (5%), roots (4%), bulbs (3%), and corms (2%), and reproductive structures like flowers and inflorescences (13%), ripe or unripe fruits (10%), and fresh seeds (4%). The majority (75%) of the cultivated vegetables have only one plant part used as a vegetable, while a quarter (25%) have multiple parts used as vegetables. Numerous vegetable species also have non-vegetable food uses such as fruit (12%) and grain/pulse (9%). *Parkia speciosa* Hassk. was the species with the most parts utilized as vegetables including the leaves, thickened inflorescences, sprouts, fruits, and seeds. Other species with many parts used as vegetables are *Moringa oleifera* Lam., *Momordica dioica* Roxb. ex Willd., *Benincasa hispida* (Thunb.) Cogn., *Sechium edule* (Jacq.) Sw., *Dioscorea praehensilis* Benth., *Nelumbo nucifera* Gaertn., *Aponogeton distachyos* L.f., *Psophocarpus grandifloras* R.Wilczek, and *Psophocarpus scandens* (Endl.) Verdc. Five groups of vegetables were defined based on their use typology: 45% are used primarily for their leaves; 19% are primarily used for underground vegetative parts (roots, tubers, corms, rhizomes, or bulbs); 8% have fruits and/or seeds used as vegetables; 7% have other vegetative parts used as vegetables such as flowers, inflorescences, stems, and shoots and 21% have multiple parts used as vegetables (Table 2).

Table 2. Use typology of cultivated vegetable species.

	Leafy Vegetables ($n = 495$)	Root Vegetables ($n = 204$)	Fruit/Seed Vegetables ($n = 90$)	Other Vegetables (Flower, Stem, Shoot) ($n = 81$)	Multiuse Vegetables ($n = 227$)
Parts used as a vegetable					
Leaves	100%				63%
Shoots, sprouts	14%			46%	25%
Stems				15%	12%
Bulb		13%			5%
Corm		9%			2%
Tuber		52%			12%
Rhizome		15%			10%
Roots		12%			11%
Flowers, petals, inflorescences				40%	48%
Fruit/pod			77%		18%
Fresh seed			30%		7%
Parts used for non-vegetable uses					
Seed	8%	1%	27%		16%
Fruit	12%	2%	20%		25%

3.2. Growth Forms

The most common growth forms of the cultivated vegetables are geophytes (33%) and therophytes (22%) (Figure 1). Phanerophytes (18%), nanophanerophytes (10%) and herbaceous phanerophytes (2%) together make up a sizable portion of the cultivated vegetables. Hemicryptophyte (6%), chamaephyte

(4%) and helophyte (4%) growth forms are less common, while only two hydrophyte (*Vallisneria natans* (Lour.) H.Hara and *Limnophila aromatic* (Lam.) Merr.) and four epiphyte (*Ficus rumphii* Blume; *Ficus annulata* Blume; *Begonia eminii* Warb.; and *Peperomia pereskiifolia* (Jacq.) Kunth) species were identified in the analysis. Of all the cultivated vegetables, 17% are climbing plants, which are mostly geophytes and therophytes. Just 3% are succulents, found mostly among the nanophanerophytes, chamaephytes, and therophytes.

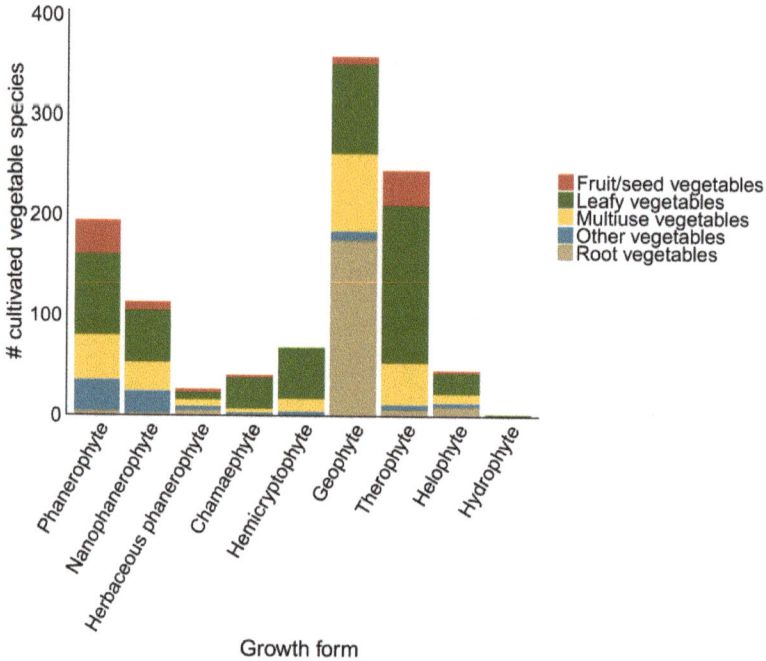

Figure 1. Growth forms of cultivated vegetable species with different uses worldwide.

3.3. Centre of Origin

The majority (72%) of cultivated vegetables have their centre of origin in just one of the world regions of crop diversity defined by Zeven and Zhukovsky [75]. Eighteen percent of the species have a wider centre of origin spanning two regions, while 10% have extensive ranges that span further than three regions. The widest ranging species include several pan-tropical (5%), Eurasian (2%), paleo-tropical (1%), and other species for which the centre of origin is unclear such as *Euphorbia hirta* L., *Neptunia oleracea* Lour., and *Laportea aestuans* (L.) Chew. Overall, 37% of cultivated vegetable species were determined to have an Asian–Pacific origin, 22% originated in the Americas, 17% are from the region spanning Europe, the Mediterranean, Near East and Central Asia, 15% originated from Saharan and sub-Saharan Africa, and 10% are wide ranging species that cross several world regions (Table S1). The geographic distribution of vegetables with different uses is shown in Figure 2.

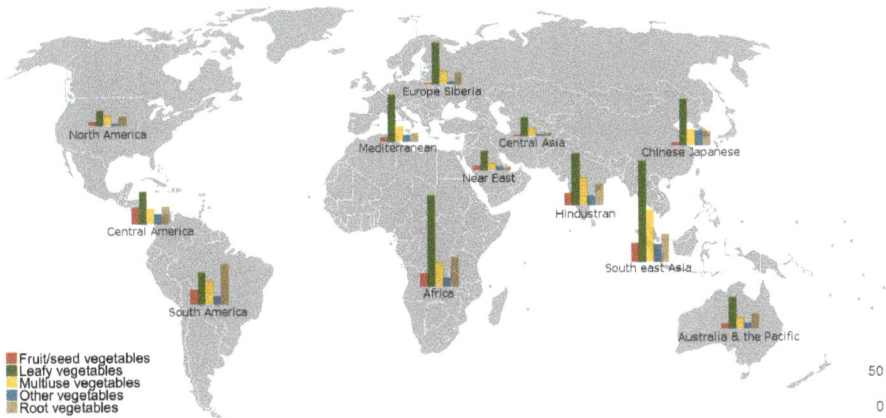

Figure 2. Number of cultivated vegetable species with different use types from different regions of diversity worldwide [75]. Base map courtesy of WikiMedia Commons.

3.4. Research

The number of Google scholar records relating to food, nutrition or vegetable uses for the cultivated vegetables ranged from 0 to 62,700 with a median of 382 (1st quartile 74; 3rd quartile 1700). No study relating to food, nutrition or vegetable uses was found for 13 of the species using the search query applied, while an additional 65 species had just 10 or fewer studies (Table S2). Many of these poorly studied vegetables were in the Dioscoreaceae ($n = 20$), Leguminosae ($n = 10$), Araceae ($n = 7$) and Compositae ($n = 5$) along with 25 other families. The best studied vegetable species were *Phaseolus vulgaris* L., *Glycine max* (L.) Merr., *Solanum lycopersicum* L., *Brassica napus* L., *Solanum tuberosum* L., *Pisum sativum* L., *Brassica oleracea* L., *Capsicum annuum* L., *Allium cepa* L., and *Vicia faba* L., each of which had more than 25,000 records in Google scholar.

The number of Google scholar records for vegetable species was significantly related to growth form, region of origin, and vegetable and non-vegetable uses (Welch's ANOVA, Table 3). Root vegetables had significantly fewer Google scholar records than all other types of vegetable (Figure 3A; Games-Howell test $p < 0.01$). The multi-use vegetables had significantly more Google scholar records than the leafy vegetables and other vegetables, as well as the root vegetables (Figure 3A; Games-Howell test $p < 0.05$). Species exclusively used as vegetables had significantly fewer Google scholar records than those with non-vegetable uses for the fruit or seed (mean 1822 ± 153 vs. 3916 ± 510; Table 3). The therophyte vegetables were by far the best researched with significantly more Google scholar records compared to chamaephyte, herbaceous phanerophyte, nanophanerophyte, phanerophyte, and geophyte vegetables (Figure 3B; Games–Howell test $p < 0.05$). Wide-ranging species and vegetables from the Europe-Mediterranean-Near East-Central Asia region had significantly more Google scholar records compared to species from Africa, the Asia-Pacific region, and the Americas (Figure 3C; Games–Howell test $p < 0.05$). Vegetables originating in Africa notably had received the lowest level of research attention, with significantly fewer Google scholar records compared to vegetables from all other regions of origin (Figure 3C; Games-Howell test $p < 0.05$).

Table 3. Results of statistical tests evaluating how indicators of neglect for cultivated vegetable species relate to growth form, region of origin, and vegetable and non-vegetable uses.

Factor	# Google Scholar Records [a]	# Accessions [a]	Documented in FAOSTAT [b]
Growth form	$F_{(7, 192.58)} = 9.84$ ***	$F_{(7, 195.5)} = 25.46$ ***	$X^2_{(7)} = 121.93$ ***
Region of origin	$F_{(4, 417.14)} = 20.94$ ***	$F_{(4, 399)} = 17.40$ ***	$X^2_{(4)} = 12.88$ *
Vegetable use	$F_{(4, 293.4)} = 18.78$ ***	$F_{(4, 299.1)} = 38.70$ ***	$X^2_{(4)} = 105.33$ ***
Non vegetable use	$F_{(1, 50.19)} = 50.19$ ***	$F_{(1, 299.1)} = 23.53$ ***	$X^2_{(1)} = 7.26$ **

* $p < 0.05$, ** $p < 0.01$, *** $p < 0.001$. [a], Welch's ANOVAs on log transformed data; [b], Chi-square tests on binary data (included as a specific species or in a group of species = 1; not included at all = 0).

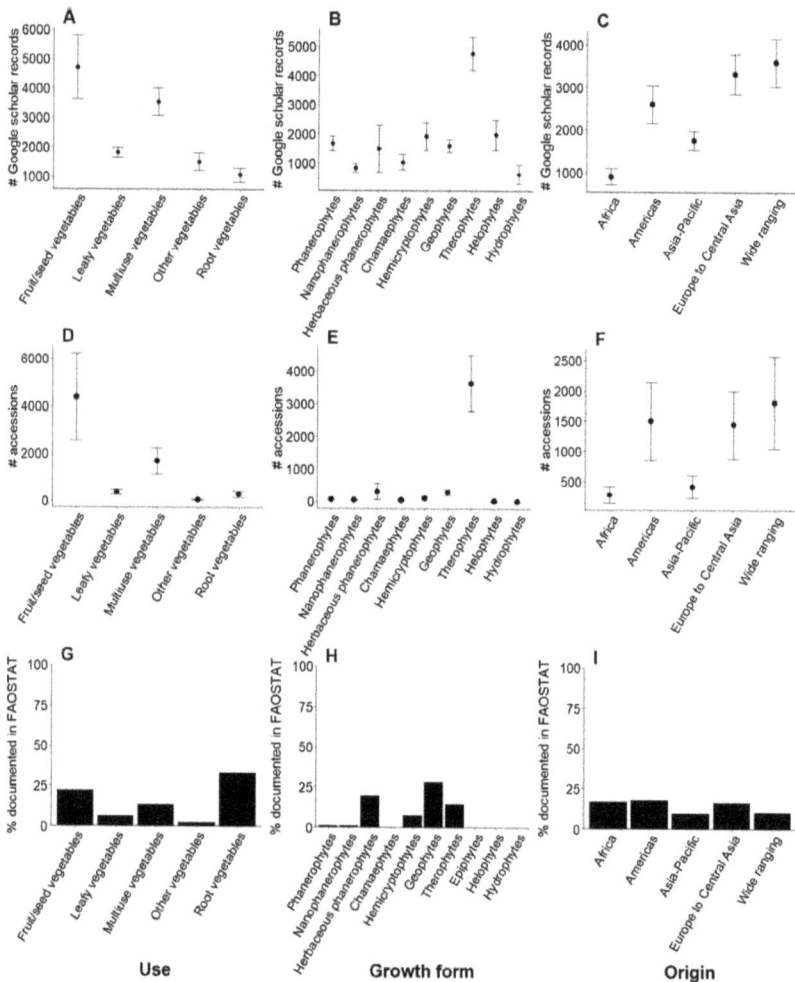

Figure 3. Mean (±standard error) number of Google scholar records relating to food, nutrition and vegetable uses for cultivated vegetables with different uses (**A**), growth forms (**B**), and regions of origin (**C**); mean (± standard error) number of accessions maintained in world genebanks for cultivated vegetables with different uses (**D**), growth forms (**E**), and regions of origin (**F**); and the percent of cultivated vegetable species with different uses (**G**), growth forms (**H**), and regions of origin (**I**) that were documented in FAOSTAT production statistics for at least one country in the previous 20 years.

3.5. Ex Situ Conservation

The number of accessions maintained for the cultivated vegetable species ranged from 0 to 142,040 with a median of 1251 accessions (1st quartile 1; 3rd quartile 50). No accessions were found to be maintained in the worlds' genebanks for 270 cultivated vegetable species (listed in Table S3). Many of the vegetables excluded from ex situ collections were in the families Dioscoreaceae ($n = 34$) and Araceae ($n = 30$) along with 79 other families. The 10 best conserved vegetable species with the most accessions were *Phaseolus vulgaris* L., *Glycine max* (L.) Merr., *Pisum sativum* L., *Solanum lycopersicum* L., *Vicia faba* L., *Capsicum annuum* L., *Solanum tuberosum* L., *Brassica oleracea* L., *Brassica napus* L. and *Allium cepa* L., which had more than 26,722 accessions each.

The number of accessions maintained by genebanks was significantly related to species' growth form, region of origin, and vegetable and non-vegetable uses (Welch's ANOVA; Table 3). Fruit/seed and multiuse vegetables had significantly more accessions than leafy vegetables, root vegetables and other vegetables (Figure 3D; Games–Howell test $p < 0.05$). Species exclusively used as vegetables had fewer genebank accessions compared to those that also had non-vegetable food uses for the fruit or seed (mean 411 ± 83 vs. 2980 ± 907; Table 3). The therophytes had significantly more accessions compared to chamaephytes, geophytes, helophytes, hemicryptophytes, phanerophytes, herbaceous phanerophytes, and nanophanerophytes (Figure 3E; Games-Howell test $p < 0.05$). The hemicryptophytes were also noted to have significantly higher numbers of accessions compared to the phanerophytes, nanophanerophytes and helophytes (Figure 3E; Games-Howell test $p < 0.05$). Vegetables from Africa and the Asia-Pacific region had significantly fewer accessions compared to species originating from the Americas and the Europe-Mediterranean-Near East-Central Asia region, as well as far-ranging species with origins spanning multiple regions (Figure 3F; Games-Howell test $p < 0.05$).

3.6. Production Data

There was a general paucity of production data in FAOSTAT for the cultivated vegetables. Only 19 species were documented specifically, while another 74 were documented in groupings that included several species, sometimes from distant taxonomic groups and including up to 20 congeners in the case of yams (Table S4). Some species could fit into multiple categories. For example, *Allium sativum* L. could be classified as "garlic" or among the "leeks and other allia". Overall, 92% of cultivated vegetable species were not covered by the database, or would only be potentially covered in very broad unspecific categories like "vegetables, fresh, not elsewhere specified" or "vegetables, leguminous, not elsewhere specified", which were excluded from our analysis for their generality. The likelihood of a species being included in FAOSTAT was significantly related to growth form, use, and region of origin (Chi-square test; Table 3). In contrast to the pattern seen for Google scholar records and genebank accessions, the root vegetables were found to have higher coverage in FAOSTAT compared to all other types of vegetable aside from fruit/seed types (Fisher exact test $p < 0.05$). The higher probability of geophytes being included in the database echoed this result (Fisher exact test $p < 0.05$). Leafy and 'other' vegetables, as well as phanerophyte and nanophanerophyte vegetables had the poorest coverage in FAOSTAT. Vegetables from the Asia-Pacific region had significantly lower probability of being included in FAOSTAT compared to those from the Americas (Fisher exact test $p < 0.05$).

4. Discussion

The results from this study confirm the existence of a large diversity of cultivated vegetables in most regions of the world, which is a rich basket of opportunities that can be harnessed to fight poverty, nutrition insecurity and vulnerability to climate change. Of the 1097 cultivated vegetables, few were found to have received substantial coverage by research, ex situ conservation, and production statistics (Table S5). Most vegetables have instead received scant attention from research and conservation efforts and their production remains poorly documented. The potential of traditional vegetables is increasingly recognized for supporting more nutritious and sustainable production and food systems [40–45], however a lack of knowledge and conservation of these species can challenge efforts for their promotion [90,91]. Clear patterns were observed regarding levels of research, conservation, and documentation of vegetables with different growth forms, uses, and regions of origin, which highlighted some priority areas to help advance the role of vegetable diversity for nutrition sensitive agriculture, as discussed in the following paragraphs.

Five use typologies of vegetables were distinguished in the study, which have received different levels of research attention. Root vegetables stood out for having significantly fewer Google scholar records compared to all other types of vegetables, calling attention to this group as potentially deserving greater research attention. The major nutritional contribution of many root vegetables is starch [1,4], but their food and nutrition security contributions can be important as they can provide important sources of health-promoting vitamins and minerals [92–94]. For example, Andean roots and tubers, such as oca (*Oxalis tuberosa* Molina) and mashua (*Tropaeolum tuberosum* Ruiz and Pav.) present distinct amino acid compositions and are rich in ascorbic acid that is fundamental for optimal absorption of iron [95]. In addition to the roots, neglected leafy vegetables and species used for stems, shoots, and flowers may also merit greater research attention as these plant parts can provide important macro and micro nutrients to diets [46–49,96,97] and they were found to have received lower research attention than the multiuse vegetables. Since more than half of cultivated vegetables (58%) are used primarily for their leaves, we note that this large and poorly studied group of species could indeed offer a great diversity of opportunities for supporting more nutrition-sensitive agriculture.

Cultivated vegetables come in a variety of growth forms including trees, shrubs, herbs, and water plants. The therophyte (annual) vegetables are by far the best researched and conserved, while other growth forms are comparatively neglected. The poor ex situ conservation of non-annual plants may relate to challenges posed by their biology and their perceived economic values. Annuals are well suited to ex situ conservation, which primarily involves storage of seeds in cold chambers [98]. Other major growth forms of vegetable such as geophytes, phanerophytes, and nanophanerophytes are often clonally propagated or have recalcitrant seeds that are sensitive to desiccation and/or cold [99,100]. The majority of plant species with recalcitrant seeds are shrubs or trees, of which about half are found in tropical moist forests [101]. Adequate representation of the genetic diversity of such species in ex situ collections poses difficulties as they must be conserved either in field genebanks or in vitro, while the processes and research required to establish their conservation may prove cost ineffective [98,102,103]. In view of these constraints, conservation of the genetic diversity of many cultivated vegetables is likely to depend in large part on in situ/on farm conservation [104]. As the use of traditional crops and transmission of associated knowledge are observed to be decreasing in many parts of the world [63,105], attention to reverse these trends are paramount to ensure the maintenance of these resources into the future.

Neglected vegetables are found in all world regions but a strikingly low amount of research and few genebank accessions are dedicated to species from Africa and the Asia-Pacific region. This pattern results from the narrow focus of research and development on major staples, as well as other historical and cultural factors that have shaped priorities in production and market development in these regions [8,106–109]. Traditional vegetables are recognized as strategic assets to reduce high rates of malnutrition that persist in Africa and Asia due to the strong nutritional values, seasonal availability, and capacity to thrive on poor soils under water limited conditions that characterize many species [46,56,110]. Important steps are already being taken for promotion of traditional vegetables towards this end [111–113]. Notably, the World Vegetable Centre (AVRDC) is conducting selection programmes for indigenous Asian and African vegetables in addition to their active breeding programmes for ten major vegetable species [114]. The African Orphan Crops Consortium is another important initiative advancing research on African crops, which is committed to developing genomics resources for 57 of the cultivated vegetable species included in this review [115]. Despite numerous important efforts such as these, very low research and conservation for African and Asian vegetables was still detected in this study, which is likely due to the vast diversity of vegetables available in these regions (406 species of vegetable in Asian Pacific Region and 165 species in Africa). Significant time, investments and policy support will be necessary to advance research, breeding, and promotion for neglected vegetables in these mega-diverse biodiversity regions, which could in turn be valuable for enabling transformations toward more nutrition sensitive agriculture.

We acknowledge that the number of Google Scholar records may not be a perfect indicator of research effort because this index cannot possibly capture all the studies that have been carried out for every species. The results were consistent with expectations that globally important crops (e.g., tomato, eggplant, cucumber, and lettuce) would have a much higher number of records compared to less common and more poorly known species, which supports the validity of this measure as an indicator of research effort. Additional indicators of research effort, such as investments in research programmes and training of researchers on specific species would have been interesting to include in the study but this information is challenging to access in a consistent and comprehensive form for global level analyses. The Agricultural Research and Development Indicators (ASTI) reveal relatively low investments in vegetables as compared to other crops and commodities in many countries (e.g., Guatemala 13% of research focused on potatoes [116]; India 8% of research focused on vegetables [117]; Mali 6% of research focused on horticultural crops [118]) but very little detail is provided in these statistics about specific crop species. Coverage in FAOSTAT is similarly not a perfect indicator of knowledge on species distribution and production levels. Much more detailed information is certainly available on the distribution and production of some species in some locations. However, accessing this data in a consistent and exhaustive form suitable for global level analyses would be very difficult. As FAOSTAT presents a standard for agricultural production statistics and is frequently relied upon by the agricultural research community for analyses of global production, we see a great value in this indicator for reflecting the level of accessible knowledge on these species. Our results highlight many gaps in the database and some peculiarities, such as higher coverage of geophytes and herbaceous phanerophytes compared to other groups that results mainly from the high number of species captured under common name categories like "yams" (20 species) or "Plantains and others" (five species). Documentation of vegetables should be vastly improved in FAOSTAT and national production statistics to support their promotion and integration into nutrition-sensitive agricultural and food systems. Disaggregating figures for different species, especially for those that are not closely related taxonomically (e.g., "Carrots and turnips"), would be an important step in this direction.

Poor documentation of production levels, as well as poor availability of data on the nutritional and agronomic characteristics of the cultivated vegetables makes it challenging to assess their use potentials. The nutritional composition of traditional vegetables is patchily documented in national and regional food composition tables [119], while the FAO EcoCrop database was found to cover only 29% of the vegetable species in our review. Among those covered, 50 species are capable of producing on low quality soil with 300 mm of rain or less annually (Table S6). These species may be relevant for supporting vegetable production in marginal areas facing climate change, however it is noted that the remaining 71% of species that are not included in this database should not be overlooked for this role, as they may also have these potentials. In this sense, generating and increasing access to information on the diversity of vegetable species can be vital toward recognizing and leveraging the potentials of cultivated vegetable species.

Many of the 1097 vegetables included in this study are neglected by research, conservation and production statistics but they may not necessarily be underutilized. Some neglected vegetable species may be popularly used in local food systems. Meanwhile others may have important limitations of toxicity, difficult processing, poor productivity, restricted growing ranges, or other constraints that could challenge efforts to promote their use [64]. For example, some of the vegetable species in our review are famine foods (e.g., *Morinda citrifolia* L., *Dioscorea sansibarensis* Pax, and *Icacina oliviformis* (Poir.) J.Raynal), which are consumed mainly in times of food shortage and have toxins that can cause unpleasant side effects such as gastrointestinal complications, demanding intensive processing to render them edible [120]. Increased research attention can help overcome key production, processing and marketing constraints to unlock their benefits for nutrition and incomes [121]. For example, traditional methods and new technologies for food processing can eradicate or reduce toxicity and antinutrients [122,123]. Breeding could also have a role in targeting changes to secondary metabolites to improve acceptability [124]. Overcoming production, processing and marketing challenges to achieve a more substantial and commercially-oriented production may not be feasible for all vegetables and may also not be efficient when alternative crops with better production and market values are available. Many of the neglected cultivated vegetables, such as those used as famine foods, may still have important roles as part of diversified landscapes and regional food systems for strengthening food security, resilience, and nutrition through all seasons and climate conditions.

Trees and shrubs that provide vegetable uses were noted in the review to have received lower attention from research and conservation compared to annual crops. These species may be highly relevant, however, for enhancing availability of nutritious foods while supporting climate change adaptation and mitigation [125,126]. Agroforestry has strong capacity for carbon sequestration and can also stabilize production in wetter and drier years thanks to the positive effects of trees on water infiltration and retention, their deep roots, and provision of alternative sources of food and income [125,127]. Agroforestry moreover provides a number of other ecosystem services, such as windbreaks, shade, structural support, fodder, and improvement of soil fertility, that reinforce farm system sustainability [128–130]. Integrating more trees into agricultural landscapes is being promoted as a climate change adaptation strategy and we note that trees and shrubs with vegetable uses could be a great fit within these approaches, while deserving greater research attention to define best practices. Previous reviews of cultivated vegetables have excluded trees and woody shrubs [70,71]. By including the woody species in this study, we propose an expanded perception of vegetables, while recognizing the potentially critical role that vegetable-providing trees and shrubs could have in climate resilient and nutrition-sensitive agroforestry systems.

This study highlights the large diversity of vegetable species that exist worldwide but it should be acknowledged that the diversity of vegetables is even greater than captured in this review. The intraspecific diversity of cultivated vegetables and the plethora of wild collected vegetables have been excluded for limitations of time and the difficulty of accessing this information. Algae and mushrooms were also excluded from the review, which include a large number of species with vegetable uses. The excluded vegetable diversity also has strategic roles for supporting more nutritious and sustainable food and farm systems and should not be overlooked. Some vegetable species have a tremendous intraspecific diversity, such as *Brassica oleracea* L. which includes important and distinct varieties such as cabbages, broccoli, cauliflower, kales, kohlrabi, collard greens, and Brussels sprouts. Different varieties can present unique tastes and features that are of tremendous cultural and culinary value and of increasing interest for marketing and improving nutrition [131,132]. Wild vegetables are also an integral and diverse component of traditional agricultural systems that continue to form a significant proportion of the global food basket [133–137]. Many wild vegetables have higher mineral and vitamin contents than cultivated vegetables [134,138,139]. In addition to these conscious exclusions from the database, it is also possible that some cultivated vegetable species have unintentionally been excluded. The species list in the Mansfeld Encyclopedia is comprehensive but it is also an evolving resource that has expanded considerably in its coverage since the first and second editions as a result of dedicated research attention and because new species have been coming into cultivation through innovations in previous decades [66].

5. Conclusions

Despite some gaps and limitations, this review has provided a good reflection of the diversity of cultivated vegetable species worldwide and trends for their research, conservation, and documentation. The study revealed that vegetables from Africa and the Asia-Pacific region have received less attention from research, conservation and production statistics as compared to vegetables from other regions, which is a gap that could be closed to leverage the role of traditional vegetables in more nutrition-sensitive agriculture in these regions. Vegetables with growth forms other than therophytes (annual plants), including many trees and shrubs with edible leaves, are largely neglected by research and conservation but merit attention to leverage their roles in agroforestry systems which can enable more sustainable vegetable production under climate change. Creating an enabling policy environment is ultimately critical for mainstreaming the use of a wider diversity of vegetables in research and development programs. Supportive policies are needed to advance research, ex situ conservation, and documentation of these species. Given the high reliance of most cultivated vegetables on in situ/on farm conservation, improving formal and informal seed systems and dissemination of relevant information to farmers (especially on cultivation requirements, resilience and nutritional benefits), strengthening the role of custodian farmers and community seed banks, increasing consumer awareness, and upgrading local value chains to encourage production are critical actions to ensure continued use and maintenance of these resources into the future. While not all 1097 cultivated vegetable species included in this study may have potential for more widespread or intensive promotion, many could have more important roles in nutrition-sensitive local production and food systems with greater attention to study, document, conserve, and promote their roles.

Supplementary Materials: The following are available online at http://www.mdpi.com/2077-0472/8/7/112/s1, Table S1: Typology of the regions of origin of cultivated vegetable species, Table S2: Cultivated vegetable species with very limited research attention related to food, vegetable or nutrition applications (1 to 10 Google scholar records), Table S3: Cultivated vegetable species with no accessions in ex situ collections, Table S4: Cultivated vegetable species with production data included and possibly included in FAOSTAT, Table S5: Cultivated vegetable species with substantial coverage by research, ex situ conservation, and production data, Table S6: Cultivated vegetable species documented in EcoCrop with capacity to produce on low quality soils with 300 mm of rain or less, Database S1: Cultivated vegetable species of the world documented in the Mansfeld Encyclopedia and their uses, growth forms and regions of origin (.csv).

Author Contributions: Conceptualization, G.M. and S.P.; Methodology, G.M. and S.P.; Data Curation, G.L., R.R., S.D. and G.M.; Investigation, G.L., R.R. and G.M.; Formal Analysis, G.M.; Visualization, G.M.; Validation, S.P., S.D. and G.M.; Writing-Original Draft Preparation, G.M., S.P., R.R. and G.L.; Writing-Review & Editing, G.M., S.P. and S.D.; Supervision, S.P. and G.M.; Funding Acquisition, S.P., Project Administration, G.M. and S.P.; Resources, Bioversity International.

Funding: This research was carried out in the framework of the project "Linking Agro biodiversity Value Chains Climate Adaptation and Nutrition: Empowering the Poor to Manage Risk" with funding from the European Commission and the International Fund for Agricultural Development (Grant 2000000978) and the CGIAR Research Programmes on Agriculture for Nutrition and Health (A4NH) and Climate Change, Agriculture and Food Security (CCAFS).

Acknowledgments: We are grateful for the constructive comments of two anonymous reviewers which helped to greatly improve the manuscript. We appreciate the support and discussions with our research team in the Healthy Diets for Sustainable Production Systems Initiative at Bioversity International, which have provided inputs and inspiration for this study. Many thanks to colleagues at FAO for supporting with FAOSTAT figures for rice, wheat and maize.

Conflicts of Interest: The authors declare no conflict of interest.

References

1. Slavin, J.L.; Lloyd, B. Health benefits of fruits and vegetables. *Adv. Nutr.* **2012**, *3*, 506–516. [CrossRef] [PubMed]

2. Liu, H.R. Health-promoting components of fruits and vegetables in the diet. *Adv. Nutr.* **2013**, *4*, 384S–392S. [CrossRef] [PubMed]

3. Herforth, A. Access to adequate nutritious food: New indicators to track progress and inform action. In *The Fight against Hunger and Malnutrition*; Sahn, D.E., Ed.; Oxford University Press: Oxford, UK, 2015.

4. World Health Organization. Healthy Diet. 2015. Available online: http://www.who.int/mediacentre/factsheets/fs394/en/ (accessed on 31 May 2018).

5. Hall, J.N.; Moore, S.; Harper, S.B.; Lynch, J.W. Global variability in fruit and vegetable consumption. *Am. J. Prev. Med.* **2009**, *36*, 402–409. [CrossRef] [PubMed]

6.   Lim, S.S.; Vos, T.; Flaxman, A.D.; Danaei, G.; Shibuya, K.; Adair-Rohani, H.; Ezzati, M. A comparative risk assessment of burden of disease and injury attributable to 67 risk factors and risk factor clusters in 21 regions. 1990–2010: A systematic analysis for the Global Burden of Disease Study 2010. *Lancet* **2012**, *380*, 2224–2260. [CrossRef]

7.   Murray, C.J.; Abraham, J.; Ali, M.K.; Alvarado, M. The state of US health. 1990–2010: Burden of diseases, injuries, and risk factors. *JAMA* **2013**, *310*, 591–608. [CrossRef] [PubMed]

8.   Pingali, P.L. Green Revolution: Impacts, limits, and the path ahead. *Proc. Natl. Acad. Sci. USA* **2012**, *109*, 12302–12308. [CrossRef] [PubMed]

9.   Kadiyala, S.; Harris, J.; Headey, D.; Yosef, S.; Gillespie, S. Agriculture and nutrition in India: Mapping evidence to pathways. *Ann. N. Y. Acad. Sci.* **2014**, *1331*, 43–56. [CrossRef] [PubMed]

10.   FAOSTAT. Production, Food Balance, and Land Use Data. Available online: http://www.fao.org/faostat/en/?#home (accessed on 18 May 2018).

11.   Schreinemachers, P.; Simmons, E.B.; Wopereis, M.C.S. Tapping the economic and nutritional power of vegetables. *Glob. Food Secur.* **2018**, *16*, 36–45. [CrossRef]

12.   Siegel, K.R.; Ali, M.K.; Srinivasiah, A.; Nugent, R.A.; Narayan, K.M.V. Do we produce enough fruits and vegetables to meet global health need? *PLoS ONE* **2014**, *9*, e104059. [CrossRef] [PubMed]

13.   Chagomoka, T.; Afari-Sefa, V.; Pitoro, R. Value chain analysis of traditional vegetables from Malawi and Mozambique. *Int. Food Agribus. Manag. Rev.* **2014**, *17*, 59–86. [CrossRef]

14.   Plazibat, I.; Ćejvanović, F.; Vasilijevic, Z. Analysis of fruit and vegetable value chains. *Bus. Excell.* **2016**, *10*, 169–189.

15.   Bandula, A.; Jayaweera, C.; De Silva, A.; Oreiley, P.; Karunarathne, A.; Malkanthi, S.H.P. Role of underutilized crop value chains in rural food and income security in Sri Lanka. *Procedia Food Sci.* **2016**, *6*, 267–270. [CrossRef]

16.   Negi, S.; Anand, N. Issues and challenges in the supply chain of fruits and vegetables sector in India: A review. *Int. J. Manag. Value Supply Chains* **2015**, *6*, 47–62. [CrossRef]

17.   Popkin, B.M. Nutrition transition and the global diabetes epidemic. *Curr. Diabetes Rep.* **2015**, *15*, 64. [CrossRef] [PubMed]

18.   Lee, A. Affordability of fruits and vegetables and dietary quality worldwide. *Lancet Glob. Health* **2016**, *4*, 664–665. [CrossRef]

19.   Miller, V.; Yusuf, S.; Chow, C.K.; Dehghan, M.; Corsi, D.J.; Lock, K.; Popkin, B.; Rangarajan, S.; Khatib, R.; Lear, S.A.; et al. Availability, affordability, and consumption of fruits and vegetables in 18 countries across income levels: Findings from the Prospective Urban Rural Epidemiology (PURE) study. *Lancet Glob. Health* **2016**, *4*, e695–e703. [CrossRef]

20.   Hawkes, C.; Harris, J.; Gillespie, S. Urbanization and the nutrition transition. *Glob. Food Policy Rep.* **2017**, *4*, 34–41. [CrossRef]

21.   U.S. Department of Agriculture. *Fresh Fruit and Vegetable Program: A Handbook for Schools*; U.S. Department of Agriculture: Washington, DC, USA, 2010. Available online: https://fns-prod.azureedge.net/sites/default/files/handbook.pdf (accessed on 15 March 2018).

22.   Carney, P.A.; Hamada, J.L.; Rdesinski, R.; Sprager, L.; Nichols, K.R.; Liu, B.Y.; Pelayo, J.; Sanchez, M.A.; Shannon, J. Impact of a community gardening project on vegetable intake, food security and family relationships: A community-based participatory research study. *J. Commun. Health* **2012**, *37*, 874–881. [CrossRef] [PubMed]

23.   Galhena, D.H.; Freed, R.; Maredia, K.M. Home gardens: A promising approach to enhance household food security and wellbeing. *Agric. Food Secur.* **2013**, *2*, 8. [CrossRef]

24.   Virchow, D.; Husmann, C.; Keatinge, J.D.H. Possibilities and constraints of horticulture for development (H4D)—An overview. *Acta Hortic.* **2016**, *1128*, 291–298. [CrossRef]

25.   Warren, E.; Hawkesworth, S.; Knai, C. Investigating the association between urban agriculture and food security. Dietary diversity, and nutritional status: A systematic literature review. *Food Policy* **2015**, *53*, 54–66. [CrossRef]

26.   Chagomoka, T.; Drescher, A.; Glaser, R.; Marschner, B.; Schlesinger, J.; Nyandoro, G. Contribution of urban and periurban agriculture to household food and nutrition security along the urban-rural continuum in Ouagadougou, Burkina Faso. *Renew. Agric. Food Syst.* **2017**, *32*, 5–20. [CrossRef]

27. Kpéra, G.N.; Segnon, A.C.; Saïdou, A.; Mensah, G.A.; Aarts, N.; van der Zijpp, A.J. Towards sustainable vegetable production around agro-pastoral dams in Northern Benin: Current situation, challenges and research avenues for sustainable production and integrated dam management. *Agric. Food Secur.* **2017**, *6*, 67. [CrossRef]

28. Singh, B.; Dwivedi, S.K. *Horticulture-based Agroforestry Systems for Improved Environmental Quality and Nutritional Security in Indian Temperate Region, Agroforestry*; Dagar, J., Tewari, V., Eds.; Springer: Singapore, 2017; pp. 245–261, ISBN 978-981-10-7650-3.

29. Weinberger, K.; Lumpkin, T.A. Diversification into horticulture and poverty reduction: A research agenda. *World Dev.* **2007**, *35*, 1464–1480. [CrossRef]

30. Springmann, M.; Mason-D'Croz, D.; Robinson, S.; Garnett, T.; Godfray, H.C.J.; Gollin, D.; Rayner, M.; Ballon, P.; Scarborough, P. Global and regional health effects of future food production under climate change: A modelling study. *Lancet* **2016**, *387*, 1937–1946. [CrossRef]

31. Tripathi, A.; Tripathi, D.K.; Chauhan, D.K.; Kumar, N.; Singh, G.S. Paradigms of climate change impacts on some major food sources of the world: A review on current knowledge and future prospects. *Agric. Ecosyst. Environ.* **2016**, *216*, 356–373. [CrossRef]

32. Snyder, R.L. Climate change impacts on water use in horticulture. *Horticulturae* **2017**, *3*, 27. [CrossRef]

33. Malholtra, S.K. Horticultural crops and climate change: A review. *Indian J. Agric. Sci.* **2017**, *87*, 12–22.

34. McDowell, J.Z.; Hess, J.J. Accessing adaptation: Multiple stressors on livelihoods in the Bolivian highlands under a changing climate. *Glob. Environ. Chang.* **2012**, *22*, 342–352. [CrossRef]

35. Quintas-Soriano, C.; Castroca, A.J.; Castroa, H.; García-Llorente, M. Impacts of land use change on ecosystem services and implications for human well-being in Spanish drylands. *Land Use Policy* **2016**, *54*, 534–548. [CrossRef]

36. Dinham, B. Growing vegetables in developing countries for local urban populations and export markets: Problems confronting small-scale producers. *Pest. Manag. Sci.* **2003**, *59*, 575–582. [CrossRef] [PubMed]

37. Ulrich, A. Export-oriented horticultural production in Laikipia, Kenya: Assessing the implications for rural livelihoods. *Sustainability* **2014**, *6*, 336–347. [CrossRef]

38. Hoi, P.V.; Mol, A.P.J.; Oosterveer, P.J.M.; van den Brink, P.J. Pesticide use in Vietnamese vegetable production: A 10-year study. *Int. J. Agric. Sustain.* **2016**, *14*, 325–338. [CrossRef]

39. Haddad, L.; Hawkes, C.; Webb, P.; Thomas, S.; Beddington, J.; Waage, J.; Flynn, D. A new global research agenda for food. *Nature* **2016**, *540*, 30–32. [CrossRef] [PubMed]

40. Ebert, A.W. Potential of underutilized traditional vegetables and legume crops to contribute to food and nutritional security, income and more sustainable production systems. *Sustainability* **2014**, *6*, 319–335. [CrossRef]

41. Chivenge, P.; Mabhaudhi, T.; Modi, A.T.; Mafongoya, P. The potential role of neglected and underutilised crop species as future crops under water scarce conditions in Sub-Saharan Africa. *Int. J. Environ. Res. Public Health* **2015**, *12*, 5685–5711. [CrossRef] [PubMed]

42. Baldermann, S.; Blagojević, L.; Frede, K.; Klopsch, R.; Neugart, S.; Neumann, A.; Ngwene, B.; Norkeweit, J.; Schröter, D.; Schröter, A.; et al. Are neglected plants the food for the future? *Crit. Rev. Plant Sci.* **2016**, *35*, 106–119. [CrossRef]

43. Sogbohossou, E.O.D.; Achigan Dako, E.G.; Maundu, P.; Solberg, S.; Deguenon, E.M.S.; Mumm, R.H.; Hale, I.; Van Deynze, A.; Schranz, M.E. A roadmap for breeding orphan leafy vegetable species: A case study of *Gynandropsis gynandra* (Cleomaceae). *Hortic. Res.* **2018**, *5*, 2. [CrossRef] [PubMed]

44. Keatinge, J.D.H.; Yang, R.Y.; Hughes, J.D.A.; Easdown, W.J.; Holmer, R. The importance of vegetables in ensuring both food and nutritional security in attainment of the millennium development goals. *Food Secur.* **2011**, *3*, 491–501. [CrossRef]

45. Nyadanu, D.; Lowor, S.T. Promoting competitiveness of neglected and underutilized crop species: Comparative analysis of nutritional composition of indigenous and exotic leafy and fruit vegetables in Ghana. *Genet. Resour. Crop Evol.* **2015**, *62*, 131–140. [CrossRef]

46. Van Jaarsveld, P.; Faber, M.; Van Heerden, I.; Wenhold, F.; van Rensburg, W.J.; Van Averbeke, W. Nutrient content of eight African leafy vegetables and their potential contribution to dietary reference intakes. *J. Food Compos. Anal.* **2014**, *33*, 77–84. [CrossRef]

47. Uusiku, N.P.; Oelofse, A.; Duodu, K.G.; Bester, M.J.; Faber, M. Nutritional value of leafy vegetables of sub-Saharan Africa and their potential contribution to human health: A review. *J. Food Compos. Anal.* **2010**, *23*, 499–509. [CrossRef]

48. Khoo, H.E.; Prasad, K.N.; Kong, K.W.; Jiang, Y.; Ismail, A. Carotenoids and their isomers: Color pigments in fruits and vegetables. *Molecules* **2011**, *16*, 1710–1738. [CrossRef] [PubMed]

49. Toledo, A.; Burlingame, B. Biodiversity and nutrition: A common path toward global food security and sustainable development. *J. Food Compos. Anal.* **2006**, *19*, 477–483. [CrossRef]

50. Maseko, I.; Mabhaudhi, T.; Tesfaym, S.; Araya, H.T.; Fezzehazion, M.; Du Plooy, C.P. African leafy vegetables: A review of status, production and utilization in South Africa. *Sustainability* **2018**, *10*, 16. [CrossRef]

51. Markus, V.; Abbey, P.A.; Yahaya, J.; Zakka, J.; Yatai, K.B.; Oladeji, M. An underexploited tropical plant with promising economic value and the window of opportunities for researchers: *Cnidoscolus aconitifolius*. *Am. J. Food Sci. Nutr. Res.* **2016**, *29*, 177.

52. Kuti, J.O.; Torres, E.S. Potential nutritional and health benefits of tree spinach. *Prog. New Crop.* **1996**, *13*, 516–520.

53. Davis, D.R.; Epp, M.D.; Riordan, H.D. Changes in USDA food composition data for 43 garden crops, 1950 to 1999. *J. Am. Coll. Nutr.* **2004**, *23*, 669–682. [CrossRef] [PubMed]

54. Davis, D.R. Declining fruit and vegetable nutrient composition: What is the evidence? *HortScience* **2009**, *44*, 15–19.

55. Weinberger, K.; Msuya, J. *Indigenous Vegetables in Tanzania—Significance and Prospects*; World Vegetable Center: Tainan, Taiwan, 2004; Volume 31, ISBN 92-9058-136-0.

56. Rubaihayo, E.B. Uganda—The Contribution Of Indigenous Vegetables to Household Food Security. Available online: https://openknowledge.worldbank.org/handle/10986/10794 (accessed on 16 May 2018).

57. Yang, R.Y.; Keding, G.B. Nutritional Contributions of Important African Indigenous Vegetables. In *African Indigenous Vegetables in Urban Agriculture*; Shackleton, C.M., Pasquini, M.W., Descher, A.W., Eds.; Earthscan: London, UK, 2009.

58. Hughes, J.D.A.; Ebert, A.W. Research and development of underutilized plant species: The role of vegetables in assuring food and nutritional security. *Acta Hortic.* **2011**, *979*, 79–92. [CrossRef]

59. Legwaila, G.M.; Mojeremane, W.; Madisa, M.E.; Mmolotsi, R.M.; Rampart, M. Potential of traditional food plants in rural household food security in Botswana. *J. Hortic. For.* **2011**, *3*, 171–177.

60. Luoh, J.; Begg, C.; Symonds, R.; Ledesma, D.; Yang, R. Nutritional yield of African indigenous vegetables in water-deficient and water-sufficient conditions. *Food Nutr. Sci.* **2014**, *5*, 812–822. [CrossRef]

61. Schiattone, M.I.; Viggiani, R.; Di Venere, D.; Sergio, L.; Cantore, V.; Todorovic, M.; Perniola, M.; Canadido, V. Impact of irrigation regime and nitrogen rate on yield, quality and water use efficiency of wild rocket under greenhouse conditions. *Sci. Hortic.* **2018**, *229*, 182–192. [CrossRef]

62. Galluzzi, G.; Lopez Noriega, I. Conservation and use of genetic resources of underutilized crops in the Americas–A continental analysis. *Sustainability* **2014**, *6*, 980–1017. [CrossRef]

63. Keller, G.B.; Mndiga, H.; Maass, B.L. Diversity and genetic erosion of traditional vegetables in Tanzania from the farmer's point of view. *Plant Genet. Resour.* **2006**, *3*, 400–413. [CrossRef]

64. Meldrum, G.; Padulosi, S. Neglected No More: Leveraging underutilized crops to address global challenges. In *Routledge Handbook of Agricultural Biodiversity*; Hunter, D., Guarino, L., Spillane, C., McKeown, P.C., Eds.; Routledge: London, UK, 2017; ISBN 9780415746922.

65. Ebert, A.W. Ex situ conservation of plant genetic resources of major vegetables. In *Conservation of Tropical Plant Species*; Normah, M.N., Chin, H.F., Reed, B.M., Eds.; Springer: New York, NY, USA, 2012; pp. 373–417. ISBN 978-1-4614-3775-8.

66. Hanelt, P.; Institute of Plant Genetics and Crop Plant Research. *Mansfeld's Encyclopedia of Agricultural and Horticultural Crops (Except Ornamentals)*, 3rd ed.; Springer: Berlin/Heidelberg, Germany, 2001; Volumes 1–6, ISBN 3540410171.

67. Watson, J.W. *Home Gardens and In Situ Conservation of Plant Genetic Resources in Farming Systems*; Bioversity International: Rome, Italy, 2002; pp. 28–29.

68. Khoshbakht, K.; Hammer, K. How many plant species are cultivated? *Genet. Resour. Crop Evol.* **2008**, *55*, 925–928. [CrossRef]

69. Khoshbakht, K.; Hammer, K. Species richness in relation to the presence of crop plants in families of higher plants. *J. Agric. R. Dev. Trop. Subtrop.* **2008**, *109*, 181–190.

70. Kayes, S.J.; Dias, J.C. Common names of commercially cultivated vegetables of the world in 15 languages. *Econ. Bot.* **1995**, *49*, 115–152. [CrossRef]

71. Rubatzky, V.E.; Yamaguchi, M. *World Vegetables: Principles, Production and Nutritive Values*, 2nd ed.; Chapman & Hall: New York, NY, USA, 1997; ISBN 978-1-4615-6015-9.

72. Radovich, J.K. *Biology and Classification of Vegetables, Handbook of Vegetables and Vegetable Processing*, 2nd ed.; Sinha, N., Hui, Y.H., Evranuz, E., Siddiq, M., Ahmed, J., Eds.; Wiley: Delhi, India, 2011; pp. 3–22, ISBN 9780470958346.

73. Kalwij, J.M. Review of 'The Plant List, a working list of all plant species'. *J. Veg. Sci.* **2012**, *23*, 998–1002. [CrossRef]

74. Cayuela, L.; Granzow-de la Cerda, Í.; Albuquerque, F.S.; Golicher, D.J. Taxonstand: An R package for species names standardisation in vegetation databases. *Methods Ecol. Evol.* **2012**, *3*, 1078–1083. [CrossRef]

75. Zeven, A.C.; Zhukovsky, P.M. *Dictionary of Cultivated Plants and Their Centres of Diversity, Excluding Ornamentals, Forest Trees and Lower Plants*; Center for Agricultural Publishing and Documentation: Wageningen, The Netherlands, 1975; pp. 1–219, ISBN 978-9022005491.

76. Raunkiaer, C. *The Life Forms of Plants and Statistical Plant Geography*; The Clarendon Press: Oxford, UK, 1934; ISBN 978-9333393362.

77. Govaerts, R.; Frodin, D.G.; Radcliffe-Smith, A. *World Checklist and Bibliography of Euphorbiaceae (with Pandanaceae)*; The Royal Botanic Gardens: Kew, UK, 2000; Volume 1, ISBN 9781900347839.

78. De Meneses Costa, A.C.; Moro, M.F.; Martins, F.R. Raunkiaerian life-forms in the Atlantic forest and comparisons of life-form spectra among Brazilian main biomes. *Braz. J. Bot.* **2016**, *39*, 833–844. [CrossRef]

79. Gour, P.G.; Sarker, A.K.; Faruq, M.O. The life-form characteristics of medicinal plants in the selected areas of Natore district, Bangladesh. *Plant Environ. Dev.* **2017**, *6*, 24–30.

80. Harzing, A.-W.; Alakangas, S. Google Scholar, Scopus and the Web of Science: A longitudinal and cross-disciplinary comparison. *Scientometrics* **2016**, *106*, 787–804. [CrossRef]

81. Shultz, M. Comparing test searches in PubMed and Google Scholar. *J. Med. Libr. Assoc.* **2007**, *95*, 442–445. [CrossRef] [PubMed]

82. Halevi, G.; Moed, H.; Bar-Ilan, J. Suitability of Google Scholar as a source of scientific information and as a source of data for scientific evaluation—Review of the literature. *J. Informetr.* **2017**, *11*, 823–834. [CrossRef]

83. Arendt, J. Imperfect tools: Google Scholar vs. Traditional commercial library databases. *Against Grain* **2008**, *17*, 20–26. [CrossRef]

84. Ramankutty, N. Croplands in West Africa: A geographically explicit dataset for use in models. *Earth Interact.* **2004**, *8*, 1–22. [CrossRef]

85. Anderson, W.; You, L.; Wood, S.; Wood-Sichra, U.; Wu, W. An analysis of methodological and spatial differences in global cropping systems models and maps. *Glob. Ecol. Biogeogr.* **2015**, *24*, 180–191. [CrossRef]

86. Kolahdooz, F.; Spearing, K.; Corriveau, A.; Sharma, S. Dietary adequacy and alcohol consumption of Inuvialuit women of child-bearing age in the Northwest Territories, Canada. *J. Hum. Nutr. Diet.* **2013**, *26*, 570–577. [CrossRef] [PubMed]

87. Järvelä-Reijonen, E.; Karhunen, L.; Sairanen, E.; Rantala, S.; Laitinen, J.; Puttonen, S.; Peuhkuri, K.; Hallikainen, M.; Juvonen, K.; Myllymäki, T.; et al. High perceived stress is associated with unfavorable eating behavior in overweight and obese Finns of working age. *Appetite* **2016**, *103*, 249–258. [CrossRef] [PubMed]

88. Games, P.A.; Howell, J.F. Pairwise multiple comparison procedures with unequal ns and or variances: A Monte Carlo study. *J. Educ. Stat.* **1976**, *1*, 113–125. [CrossRef]

89. Day, R.W.; Quinn, G.P. Comparisons of treatments after an analysis of variance in ecology. *Ecol. Monogr.* **1989**, *59*, 433–463. [CrossRef]

90. Mambolco, T.F. Nutrients and Antinutritional Factors at Different Maturity Stages of Selected Indigenous African Green Leafy Vegetables. Ph.D. Thesis, Sokoine University of Agriculture, Morogoro, Tanzania, 2015.

91. Mnzava, N.A. Vegetable crop diversification and the place of traditional species in the tropics, traditional African vegetables. In *Promoting the Conservation and Use of Underutilized and Neglected Crops*; Guarino, L., Ed.; Institute of Plant Genetics and Crop Plant Research, Gatersleben/International Plant Genetic Resources Institute: Rome, Italy, 1997.

92. Hotz, C.; Loechl, C.; de Brauw, A.; Eozenou, P.; Gilligan, D.; Moursi, M.; Munhaua, B.; van Jaarsveld, P.; Carriquiry, A.; Meenakshi, J.V. A large-scale intervention to introduce orange sweet potato in rural Mozambique increases vitamin A intakes among children and women. *Br. J. Nutr.* **2012**, *108*, 163–176. [CrossRef] [PubMed]

93. Devaux, A.; Kromann, P.; Ortiz, O. Potatoes for sustainable global food security. *Potato Res.* **2014**, *57*, 185–199. [CrossRef]

94. Ferraro, V.; Piccirillo, C.; Tomlins, K.; Pintado, M.E. Cassava (*Manihot esculenta* Crantz) and yam (*Dioscorea* spp.) crops and their derived foodstuffs: safety, security and nutritional value. *Crit. Rev. Food Sci. Nutr.* **2015**, *56*, 2714–2727. [CrossRef] [PubMed]

95. Flores, H.E.; Walker, T.S.; Guimarães, R.L.; Bais, H.P.; Vivanco, J.M. Andean root and tuber crops: Underground rainbows. *HortScience* **2003**, *38*, 161–167.

96. Chongtham, N.; Bisht, M.S.; Haorongbam, S. Nutritional properties of bamboo shoots: Potential and prospects for utilization as a health food. *Compr. Rev. Food Sci. Food Saf.* **2011**, *10*, 153–168. [CrossRef]

97. Rop, O.; Mlcek, J.; Jurikova, T.; Neugebauerova, J.; Vabkova, J. Edible flowers—A promising source of mineral elements in human nutrition. *Molecules* **2012**, *17*, 6672–6683. [CrossRef] [PubMed]

98. Cruz-Cruz, C.A.; González-Arnao, M.T.; Engelmann, F. Biotechnology and conservation of plant biodiversity. *Resources* **2013**, *2*, 73–95. [CrossRef]

99. Dulloo, M.E.; Hunter, D.; Borelli, T. Ex situ and in situ conservation of agricultural biodiversity: Major advances and research needs. *Not. Bot. Hort. Agrobot. Cluj Napoca* **2010**, *38*, 123–135. [CrossRef]

100. McKey, D.; Elias, M.; Pujol, B.; Duputié, A. The evolutionary ecology of clonally propagated domesticated plants. *New Phytol.* **2010**, *186*, 318–332. [CrossRef] [PubMed]

101. Tweddle, J.C.; Dickie, J.B.; Baskin, C.C.; Baskin, J.M. Ecological aspects of seed desiccation sensitivity. *J. Ecol.* **2003**, *91*, 294–304. [CrossRef]

102. Dawson, I.K.; Guariguata, M.R.; Loo, J.; Weber, J.C.; Lengkeek, A.; Bush, D.; Cornelius, J.; Guarino, L.; Kindt, R.; Orwa, C.; et al. What is the relevance of smallholders' agroforestry systems for conserving tropical tree species and genetic diversity in circa situm, in situ and ex situ settings? *Biodivers. Conserv.* **2013**, *22*, 301–324. [CrossRef]

103. Walters, C.; Berjak, P.; Pammenter, N.; Kennedy, K.; Raven, P. Preservation of recalcitrant seeds. *Science* **2013**, *339*, 915–916. [CrossRef] [PubMed]

104. Fowler, C.; Hodgkin, T. Plant genetic resources for food and agriculture: assessing global availability. *Annu. Rev. Environ. Resour.* **2004**, *29*, 143–179. [CrossRef]

105. Chorol, S.; Angchok, D.; Angmo, P.; Tamchos, T.; Singh, R.K. Traditional knowledge and heirloom root vegetables: Food security in trans-Himalayan Ladakh, India. *Indian J. Tradit. Knowl.* **2018**, *17*, 191–197.

106. Evenson, R.E.; Gollin, D. Assessing the impact of the green revolution, 1960 to 2000. *Science* **2003**, *300*, 758–762. [CrossRef] [PubMed]

107. National Research Council. *Lost Crops of Africa: Vegetables*; The National Academies Press: Washington, DC, USA, 2006; Volume 2, ISBN 978-0-309-16454-2.

108. Arora, R.K. *Diversity in Underutilized Plant Species: An Asia-Pacific Perspective*; Bioversity International: New Delhi, India, 2014; ISBN 78-92-9255-007-3.

109. Heady, D.; Hoddinott, J. Agriculture, nutrition and the green revolution in Bangladesh. *Agric. Syst.* **2016**, *149*, 122–131. [CrossRef]

110. Kamga, R.T.; Kouamé, C.; Atangana, A.R.; Chagomoka, T.; Ndango, R. Nutritional evaluation of five African indigenous vegetables. *J. Hortic. Res.* **2013**, *21*, 99–106. [CrossRef]

111. Oluoch, M.O.; Pichop, G.N.; Silué, D.; Abukutsa-Onyango, M.O.; Diouf, M.; Shackleton, C.M. Production and Harvesting Systems for African INDIGENOUS VEGETABLES. In *African Indigenous Vegetables in Urban Agriculture*; Shackleton, C.M., Pasquini, M.W., Descher, A.W., Eds.; Earthscan: London, UK, 2009.

112. Gotor, E.; Irungu, C. The impact of Bioversity International's African Leafy Vegetables programme in Kenya. *Impact Assess. Proj. Apprais.* **2012**, *28*, 41–55. [CrossRef]

113. Food and Agriculture Organization. *Future Smart Food: Rediscovering Hidden Treasures of Neglected and Underutilized Species for Zero Hunger in Asia*; Food and Agriculture Organization of the United Nations: Bangkok, Thailand, 2018; Available online: http://www.fao.org/3/I8907EN/i8907en.pdf (accessed on 31 May 2018).

114. World Vegetable Center. Vegetable Diversity and Improvement. Available online: https://avrdc.org/our-work/developing-new-varieties/ (accessed on 25 May 2018).

115. African Orphan Crops Consortium. Meet the Crops. Available online: http://africanorphancrops.org/meet-the-crops/ (accessed on 25 May 2018).

116. Perez, S.; Martínez, J.; Beintema, N.; Flaherty, K. Agricultural R&D Indicators Factsheet Guatemala. Available online: https://www.asti.cgiar.org/pdf/factsheets/Guatemala-Factsheet.pdf (accessed on 29 June 2018).

117. Stads, G.J.; Sastry, K.; Kumar, G.; Kondisetty, T.; Gao, L. Agricultural R&D Indicators Factsheet India. Available online: https://www.asti.cgiar.org/sites/default/files/pdf/factsheets/India-Factsheet.pdf (accessed on 29 June 2018).

118. Magne Domgho, L.V.; Traoré, O.; Stads, G.J. Agricultural R&D Indicators Factsheet Mali. Available online: https://www.asti.cgiar.org/sites/default/files/pdf/Mali-Factsheet-2017.pdf (accessed on 29 June 2018).

119. Schönfeldt, H.C.; Pretorius, B. The nutrient content of five traditional South African dark green leafy vegetables—A preliminary study. *J. Food Compos. Anal.* **2011**, *24*, 1141–1146. [CrossRef]

120. Guinand, Y.; Lemessa, D. Wild-food plants in Ethiopia: Reflections on the role of wild foods and famine foods at a time of drought. *Potential Indig. Wild Foods* **2001**, *22*, 39.

121. Padulosi, S.; Amaya, K.; Jäger, M.; Gotor, E.; Rojas, W.; Valdivia, R. A Holistic approach to enhance the use of neglected and underutilized species: The case of Andean grains in Bolivia and Peru. *Sustainability* **2014**, *6*, 1283–1312. [CrossRef]

122. Terangpi, R.; Ratan Basumatary, R.T. Nutritional consideration of three important emergency food plants studied among Karbi Tribe of North East India. *J. Sci. Innov. Res.* **2015**, *4*, 138–141.

123. Getachew, A.; Asfaw, Z.; Singh, V.; Woldu, Z.; Baidu-Forson, J.J.; Bhattacharya, S. Dietary values of wild and semi-wild edible plants in Southern Ethiopia. *Afr. J. Food Agric. Nutr. Dev.* **2013**, *13*. Available online: https://www.ajol.info/index.php/ajfand/article/view/87478 (accessed on 31 May 2018).

124. Meyer, R.S.; DuVal, A.E.; Jensen, H.R. Patterns and processes in crop domestication: An historical review and quantitative analysis of 203 global food crops. *New Phytol.* **2012**, *196*, 29–48. [CrossRef] [PubMed]

125. Verchot, L.V.; Van Noordwijk, M.; Kandji, S.; Tomich, T.; Ong, C.; Albrecht, A.; Bantilan, M.C.; Anupama, K.V.; Palm, C.J. Climate change: Linking adaptation and mitigation through agroforestry. *Mitig. Adapt. Strat. Glob. Chang.* **2007**, *12*, 901–918. [CrossRef]

126. Mbow, C.; Smith, P.; Skole, D.; Duguma, L.; Bustamante, M. Achieving mitigation and adaptation to climate change through sustainable agroforestry practices in Africa. *Curr. Opin. Environ. Sustain.* **2014**, *6*, 8–14. [CrossRef]

127. Thorlakson, T.; Neufeldt, H. Reducing subsistence farmers' vulnerability to climate change: Evaluating the potential contributions of agroforestry in western Kenya. *Agric. Food Secur.* **2012**, *1*, 15. [CrossRef]

128. Jose, S. Agroforestry for ecosystem services and environmental benefits: An overview. *Agrofor. Syst.* **2009**, *76*, 1–10. [CrossRef]

129. Sileshi, G.W.; Debusho, L.K.; Akinnifesi, F.K. Can integration of legume trees increase yield stability in rainfed maize cropping systems in Southern Africa? *Agron. J.* **2012**, *104*, 1392–1398. [CrossRef]

130. Asbjornsen, H.; Hernandez-Santana, V.; Liebman, M.; Bayala, J.; Chen, J.; Helmers, M.; Schulte, L. Targeting perennial vegetation in agricultural landscapes for enhancing ecosystem services. *Renew. Agric. Food Syst.* **2014**, *29*, 101–125. [CrossRef]

131. Elia, A.; Santamaria, P. Biodiversity in vegetable crops, a heritage to save: The case of Puglia region. *Ital. J. Agron.* **2013**, *8*, 4. [CrossRef]

132. Hurtado, M.; Vilanova, S.; Plazas, M.; Gramazio, P.; Herraiz, F.J.; Andújar, I.; Prohens, J.; Castro, A. Enhancing conservation and use of local vegetable landraces: The Almagro eggplant (*Solanum melongena* L.) case study. *Genet. Resour. Crop Evol.* **2014**, *61*, 787–795. [CrossRef]

133. Crivetti, L.E.; Ogle, B.M. Value of traditional foods in meeting macro- and micronutrient needs: The wild plant connection. *Nutr. Res. Rev.* **2000**, *13*, 31–46. [CrossRef] [PubMed]

134. Flyman, M.V.; Afolayan, A.J. The suitability of wild vegetables for alleviating human dietary deficiencies. *S. Afr. J. Bot.* **2006**, *72*, 492–497. [CrossRef]

135. Bharucha, Z.; Pretty, J. The roles and values of wild foods in agricultural systems. *Philos. Trans. R. Soc. Lond. B Biol. Sci.* **2010**, *365*, 2913–2926. [CrossRef] [PubMed]

136. Sánchez-Mata, M.C.; Cabrera Loera, R.D.; Morales, P.; Fernández-Ruiz, V.; Cámara, M.; Díez Marqués, C.; Pardo-de-Santayana, M.; Tardío, J. Wild vegetables of the Mediterranean area as valuable sources of bioactive compounds. *Genet. Resour. Crop Evol.* **2012**, *59*, 431–443. [CrossRef]
137. Salvi, J.; Katewa, S.S. A review: Underutilized wild edible plants as a potential source of alternative nutrition. *Int. J. Bot. Stud.* **2016**, *1*, 32–36.
138. Afolayan, A.J.; Jimoh, F.O. Nutritional quality of some wild leafy vegetables in South Africa. *Int. J. Food Sci. Nutr.* **2009**, *60*, 424–431. [CrossRef] [PubMed]
139. Vorster, I.H.J.; van Rensburg, W.J.; Venter, S.L. The importance of traditional leafy vegetables in South Africa. *Afr. J. Food Agric. Nutr. Dev.* **2007**, *7*, 1–13.

agriculture

MDPI

Article

BiodiverSO: A Case Study of Integrated Project to Preserve the Biodiversity of Vegetable Crops in Puglia (Southern Italy)

Massimiliano Renna [1,2], Francesco F. Montesano [2,*], Angelo Signore [1,*], Maria Gonnella [2] and Pietro Santamaria [1]

[1] Department of Agricultural and Environmental Science, University of Bari Aldo Moro, Via Amendola 165/A, Bari, Italy; massimiliano.renna@uniba.it (M.R.); pietro.santamaria@uniba.it (P.S.)

[2] Institute of Sciences of Food Production (ISPA), CNR, Via Amendola 122/O, Bari, Italy; maria.gonnella@ispa.cnr.it

* Correspondence: francesco.montesano@ispa.cnr.it (F.F.M.); angelo.signore@uniba.it (A.S.); Tel.: +39-080-544-3098

Received: 29 June 2018; Accepted: 16 August 2018; Published: 18 August 2018

Abstract: Puglia region is particularly rich in agro-biodiversity, representing an example of how local vegetables varieties can still strongly interact with modern horticulture. Unfortunately, the genetic diversity of vegetable crops in this region has been eroded, due to several factors such as abandonment of rural areas, ageing of the farming population, and failure to pass information down the generations. This article summarizes the objectives, methodological approach and results of the project "Biodiversity of the Puglia's vegetable crops (BiodiverSO)", an integrated project funded by Puglia Region Administration under the 2007–2013 and 2014–2020 Rural Development Program (RDP). Results were reported for each of the eight activities of the project. Moreover, the Polignano carrot (a local variety of *Daucus carota* L.) was described as a case study, since several tasks have been performed within all eight project activities with the aim of verifying the effectiveness of these actions in terms of safeguarding for this genetic resource strongly linked with local traditions. BiodiverSO is an example of protection and recovery of vegetables at risk of genetic erosion that could help to identify and valorize much of the Puglia's plant germplasm.

Keywords: characterization; conservation; databases; genetic resources; history; local varieties; recovery; sanitation; seed bank

1. Introduction

The concepts of biodiversity, of its progressive loss and the need to protect it, are now rooted in the scientific and policy maker communities. These concepts are also increasingly gaining popularity with ordinary people. According to a survey on attitudes of Europeans toward biodiversity, at least eight out of ten Europeans consider the various effects of biodiversity loss to be serious. However, although the majority of Europeans have heard of the term "biodiversity" (60%), less than one third (30%) know what it means, most do not feel informed about biodiversity loss (66%), and think the EU should better inform citizens about the importance of biodiversity (93%) [1].

In the general framework of biodiversity issue, which is in most cases referred to natural ecosystems, the concept of "agro-biodiversity" is arising lively interest. According to the FAO definition [2], agro-biodiversity is a vital sub-set of biodiversity, and refers to the diversity in agro-ecosystems. It comprises the diversity of living organisms and genetic resources (cultivated species, varieties, and breeds; wild flora; soil microorganisms; predators; and pollinators). Agro-biodiversity is the result of interaction among the environment, genetic resources and management systems and practices, encompassing the variety and variability that are necessary for sustaining food production and food security. Therefore, local knowledge and culture can be considered as integral parts of agro-biodiversity, because the human activity of agriculture shapes and conserves this biodiversity.

Intensive agriculture has generally resulted in higher productivity, but also in a trend towards decreasing levels of agro-biodiversity. It has been noted that the so called "green revolution" in agriculture, with its modern scientific approaches to plant breeding, represented a biodiversity narrowing phase, by replacing genetically diverse landraces and local varieties (selected over centuries and representing an incredible heritage of diversity) with uniform varieties such as hybrid F1 in vegetables [3,4]. As reported by some authors [5], *"a local variety (also called: landrace, farmer's variety, folk variety) is a population of seed- or vegetative-propagated crop characterized by greater or lesser genetic variation, which is however well identifiable and which usually has a local name ... has not been subjected to an organized program of genetic improvement ... is characterized by a specific adaptation to the environmental and cultivation conditions of the area where it has been selected ... is closely associated with the traditions, the knowledge, the habits, the dialects and the occurrences of the human population that have developed it and/or continue its cultivation"*. During the last century, almost 75% of local varieties have been lost, but this percentage may rise to 90% in USA. Modern varieties are conceived to meet the requirements of market, processing industry and modern distribution. At the same time, they are subjected to rapid obsolescence.

The preservation of agro-biodiversity represents a key-point to assure adaptability and resilience of agro-ecosystems to the global challenge we will be facing in the near future to produce more and better food in a sustainable way. However, many components of agro-biodiversity would not survive without human interference. On the other hand, human choices may represent a threat for the agro-biodiversity preservation.

Puglia region is located in the southeastern part of Italy. It is largely open to the Adriatic and Ionian seas with a coastal zone of nearly 800 km. The area of about 19,360 km^2 shows more than 60% of territory below 200 m above sea level, with some peaks of more than 1000 m located in the northeast and northwest. Puglia has a typically Mediterranean climate with temperatures that may fall below 0 °C in winter (in the northern part or hills) and exceed 40 °C in summer. Annual rainfall ranges between 400 and 550 mm, mostly concentrated during the winter.

Due to its climatic conditions and land characteristics, Puglia is one of the most important regions in Italy for the vegetable production, accounting for ≈21% (≈92,000 ha) and ≈24% (≈3,261,000 tons) of the total open air growing area and the amount of vegetables produced at national level, respectively [6]. The region is among the leaders for the production of several vegetable crops such as broccoli and cauliflower, celery, parsley, processing tomato, artichoke, endive and escarole, cabbage, fennel, lettuce, cucumber, early potato and asparagus. The vegetable production industry accounts for about 30% of the total economic value of the regional agricultural sector. About 8000 ha of the regional vegetable growing area are interested by organic cultivation systems, representing the 30% of the total national organic vegetable cultivation [7].

At the same time, at national level, the predominant position of Puglia in terms of vegetable production industry does not overlap at all with its position in the seed industry (less than 2% of commercial vegetable varieties registered in the national register are from companies located in the region). However, Puglia represents an example of how local vegetable varieties can still strongly interact with modern horticulture in the definition of a complex food system, in which local culture and traditions are interconnected with local productions and local environment. In fact, the region is particularly rich in local varieties of vegetables. Some of them are still largely used and requested by the population.

Unfortunately, the genetic diversity of vegetable crops in the Puglia region has been eroded, due to several factors such as abandonment of rural areas, ageing of the farming population, and failure to pass information down the generations (leading to loss of knowledge and historical memory), which can vary in relation to the type of genetic resource and location [8,9]. Moreover, loss of family vegetable gardens near settlements, as a consequence of intense urban development, has led to a considerable loss of agro-biodiversity, especially among folk varieties grown for family consumption [4].

For these reasons, the Puglia Region Administration planned two specific actions under the 2007–2013 and 2014–2020 Rural Development Programme (RDP) to preserve regional genetic biodiversity. "Protection of biodiversity" provides financial support for a five-year period for seed savers committed to preserving in situ the plant genetic resources listed in a specific annex of the RDP, while "Integrated projects and regional biodiversity system" funds the salvage of native plant genetic resources and knowledge of ethno-botany. The overall goal is to create a biodiversity network to promote the exchange of information between stakeholders to facilitate the diffusion and protection of genetic resources in agriculture.

This article aims to pointing out objectives, methodological approach and results of the project "Biodiversity of the Puglia's vegetable crops (BiodiverSO)", one of the five integrated project funded by Puglia Region Administration under RDP.

2. Methodology

BiodiverSO (https://biodiversitapuglia.it/) is a collaborative project funded within the RDP of the Puglia Region Administration on the topic "Diffusion and protection of genetic resources in agriculture". The 48-month project started in April 2013 involving fourteen partners (https://biodiversitapuglia.it/partner/) from academia, research institutions, public consortia and private companies from different disciplines (genetic, agriculture, biology, biotechnology, chemistry, biochemistry, plant science and engineering). The project was coordinated by the Department of Agricultural and Environmental Science, University of Bari Aldo Moro.

According to the aims of BiodiverSO, project activities were organized and scheduled as eight different Work Packages regarding the local varieties of vegetables in Puglia.

WP 1—History. This activity consisted in acquiring information on local genetic resources, through an accurate bibliographic and territorial investigation work. The aim was to retrieve the history, provenance, distribution on the territory, knowledge and traditions related to cultivation and use of local varieties.

WP 2—Recovery. The recovery activity was based on research and subsequent collection of propagation material of the local varieties present in the Puglia territory. Seeds or other propagation material were taken from the field, with the relevant information on the cultivation technique, uses of the products and local traditions.

WP 3 Ex situ conservation. This activity involved the conservation, outside the natural environment (ex situ), of seeds and/or plant parts suitable for multiplication, sowing and/or planting, revitalization, propagation and management of the material, to ensure the use of germplasm, safeguard varieties at risk of genetic extinction and attempt a subsequent reintroduction.

WP 4 Characterization. This activity aimed to assess genetic, agronomic, morphological, quantitative and nutritional traits of the recovered genetic resources by using several tools and techniques.

WP 5—Sanitation and registration in national catalogs. The first aim of this activity was to obtain protocols for producing virus-free genetic resources for conservation, breeding, and production. Moreover, the registration of local varieties and related products in the "List of Traditional Agri-Food Products" (LTAFP) of the Italian Department for Agriculture was carried out.

WP 6—Databases. All information and results obtained through each project activity were cataloged and used to create inventories regarding sampling sites and data on the recovered genetic resources.

WP 7—Data sheets writing. This activity was carried out for each genetic resource with the aim to create standardized systems of identification, characterization and recognition of local varieties.

WP 8—In situ conservation. In situ conservation of the recovered propagation material was carried out through reproduction and cultivation, in isolation and in conditions of maximum purity, within local farms.

3. Obtained Results and Discussion

3.1. Project's General Results

3.1.1. WP 1—History

The information obtained from bibliographic and territorial investigations allowed obtaining a multimedia library that collects, makes accessible and available the knowledge acquired on local genetic resources (https://biodiversitapuglia.it/biblioteca/?idt=1). Thanks to this activity, 243 documents were recovered (Figure 1). Moreover, interviews, videos, (https://biodiversitapuglia.it/biblioteca/?idt=4) geo-referenced photos and 125 thematic maps (https://biodiversitapuglia.it/biblioteca/?idt=6) were produced.

Figure 1. Examples of ancient documents recovered under the bibliographic research activity.

Very interesting information on local varieties was recovered during this activity. For example, the "Lucera artichoke", a landrace of *Cynara cardunculus* L. subsp. *scolymus* Hayek, has been cultivated for decades by local smallholder farmers in Capitanata (Puglia, Italy) around the town of Lucera (Foggia). Its flower heads have an ovoid shape, while the color of the external bracts is green with purple nuance. The length is about 10 cm, the diameter 7.5–8.5 cm and the weight ranges 150–180 g (Figure 2).

Figure 2. A typical flower head of the "Lucera artichoke".

At the beginning of the 1900s, this landrace of artichoke was cultivated in many family gardens in the urban surroundings of Lucera. Unfortunately, these areas have been affected by intense building activity that has irreversibly subtracted fields traditionally used for horticulture. This has determined a considerable reduction of the "Lucera artichoke" production as well as a loss of knowledge and historical memory, and today this landrace is grown In only a few home gardens by some old farmers.

The historical importance of the "Lucera artichoke" is due to the presence of the Saracens in Puglia during the 12th century. In effect, Egidi [10] reported that a colony of Saracens was transferred into Lucera city between 1224 and 1225 A.C. by King Federico II of Svevia. Furthermore, investigations on historical texts have been carried out about Saracens customs during ancient times in Puglia. Findings indicate that the cultivation of *kinaria* (artichoke) was practiced by the Saracens in Southern Italy at least since the 12th century. Thus, it is possible that the cultivation of the artichoke began at least two centuries before with respect to what is generally reported in several publications dealing with horticulture. In this context, further study could be carried out with the aim to verify if the *kinaria* cultivated by Saracens in Capitanata during ancient times represents a possible ancestor of the "Lucera artichoke" landrace.

3.1.2. WP 2—Recovery

Several companies, farmers or simple enthusiasts that conserve and cultivate local vegetable varieties at risk of genetic erosion were identified. Seeds (Figure 3) or other propagation material have been taken from the field together to relevant information on the cultivation technique, uses of the products and local traditions. In addition, geographic and topographical information on the cultivation areas of local varieties have been acquired. Thanks to this activity, about 350 companies have been visited, over 530 accessions have been recovered and the relative descriptive sheets have been prepared. The very important role of home gardens as repositories of agro-diversity has been well verified. Home gardens, whether found in rural or peri-urban areas, are characterized by a structural complexity and multi-functionality which enables the provision of different benefits to ecosystems and people. At the same time, home gardens are important social and cultural spaces where knowledge related to agricultural practices is transmitted. Therefore, it is very important to promote their role in the conservation of vegetables crops and cultural heritage.

Figure 3. Example of accessions seeds recovered from farmers or simple enthusiasts during the WP2 activity (Photo credit: Beniamino Leoni).

3.1.3. WP 3—Ex Situ Conservation

Ex situ conservation of the recovered propagation material was carried out through the creation and management of catalog fields within experimental farms (Figure 4), long-time seeds preservation in seed banks (Figure 5) as well as the maintenance of propagation material by in vitro and/or slow-growing techniques (Figure 6), aimed to keep the agamic propagation material in good conditions, reducing conservation costs and space.

Figure 4. Catalog fields of *Brassicaeae* (**left**) and different local varieties of *Solanum lycopersicum* L. (**right**) within "La Noria" experimental farm of the National Research Council, Bari, Italy.

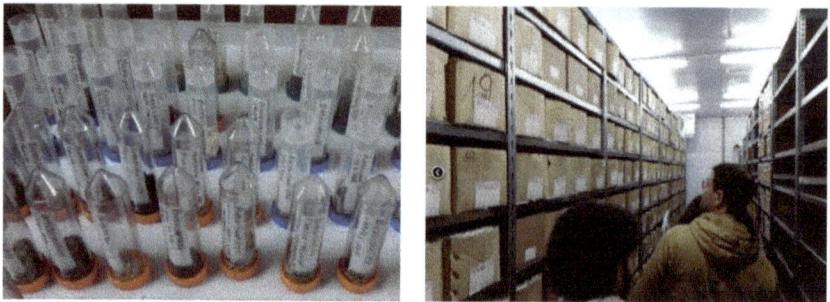

Figure 5. Example of accessions seeds (**left**) recovered to be preserved at the seed bank (**right**) of the Institute of Biosciences and Bioresources of the Italian National Research Council in Bari, Italy (Photo credit: Beniamino Leoni).

Figure 6. Shoots of *Ipomoea batatas* L. after 12 months of in vitro slow-growing (Photo credit: Claudia Ruta—from https://biodiversitapuglia.it).

3.1.4. WP 4—Characterization

Several local varieties and wild edible plants (a total of 141 genotypes) were characterized for morphological, chemical (including qualitative and nutritional) and genetic traits (Table 1).

Table 1. Genetic resources studied during the project activity (adapted from Accogli et al. [11]). For each one, performed activities are indicated by using X symbol or a specific acronym. Registration column reports the state of art regarding Puglia local varieties with Protected Geographical Indication or registered as an item in the List of "Traditional Agri-Food Products" of the Italian Department for Agriculture.

Taxonomy	Local Variety Name	Characterization			Sanitation	Registration
		M	C	G		
Allium cepa L.	Cipolla bianca di Margherita	X	X			PGI
Allium cepa L.	Cipolla rosa o dorata di Monteleone	GIBA [1]	X			
Allium cepa L.	Cipolla rossa di Acquaviva	X	X	X		TAFP
Allium cepa L.	Cipolla rossa di Margherita	GIBA [1]	X			
Allium cepa L.	Sponzale rosso di Acquaviva	X	X	X		TAFP
Allium sativum L.	Aglio dei Cortigli	X	X			
Allium sativum L.	Aglio del Salento	X				
Allium sativum L.	Aglio di Altamura	X	X			
Allium sativum L.	Aglio di Anzano	X	X			
Allium sativum L.	Aglio di Peschici	X	X			
Allium sativum L.	Aglio rosso di Monteleone e Panni	X	X			
Apium graveolens L. group *dulce*	Sedano di Torrepaduli	GIBA [2]				
Beta vulgaris L. subsp. *vulgaris* L.	Bietola da costa barese	GIBA [3]	X	X		
Beta vulgaris L. subsp. *vulgaris* L.	Bietola da costa di Fasano	GIBA [3]	X	X		
Brassica oleracea L. group *acephala*	Cavolo riccio	GIBA [4]	X	X		TAFP
Brassica oleracea L. group *botrytis*	Cima di cola	GIBA [5]	X	X		TAFP
Brassica oleracea L. group *capitata*	Cavolo cappuccio cuore di bue	GIBA [6]	X			
Brassica oleracea L. group *gongylodes*	Testa di morto	X		X		
Brassica oleracea L. group *italica*	Cima nera	GIBA [7]	X	X		
Brassica oleracea L. group *italica*	Mugnolicchio	X				
Brassica oleracea L. group *italica*	Mùgnulu	GIBA [7]	X	X		TAFP
Brassica oleracea L. group *italica*	Turzella	X				
Brassica rapa L. group *broccoletto*	Cima di rapa di montagna	X	X			
Brassica rapa L. group *broccoletto*	Cima di rapa quarantina	X	X	X		TAFP
Brassica rapa L. group *broccoletto*	Cima di rapa sessantina	X	X	X		TAFP
Brassica rapa L. group *broccoletto*	Cima di rapa novantina	X	X	X		TAFP
Brassica rapa L. group *broccoletto*	Cima di rapa centoventina	X		X		TAFP
Capsicum annuum L.	Peperone corna di capra	X	X			
Capsicum annuum L.	Peperone cornaletto paesano	X				
Capsicum annuum L.	Peperone cornaletto riccio	GIBA [8]				
Capsicum annuum L.	Peperone cornetto bianco	GIBA [8]				
Capsicum annuum L.	Peperone cornetto di Gravina	X	X			
Capsicum annuum L.	Peperone cornetto di Palagianello	X	X			
Capsicum annuum L.	Peperone cornetto (per friscere) di Presicce	X	X			
Capsicum annuum L.	Peperone cornetto leccese	X				
Capsicum annuum L.	Peperone corno di toro	GIBA [8]				
Capsicum annuum L.	Peperone "diavolicchio" di Gravina	X	X			
Capsicum annuum L.	"Papecchia"	X	X			
Capsicum annuum L.	"Paperule a core"	X	X			
Capsicum annuum L.	"Papirussi a cornulara" di Zollino	X	X			
Capsicum annuum L.	Peperone tondo rosso "a cumbost"	X	X			
Capsicum annuum L.	Peperone zanzari	GIBA [8]	X			
Cichorium intybus L. group *catalogna*	Cicoria bianca di Tricase	X	X			
Cichorium intybus L. group *catalogna*	Cicoria di Brindisi	X	X			
Cichorium intybus L. group *catalogna*	Cicoria di Galatina	X	X			TAFP
Cichorium intybus L. group *catalogna*	Cicoria di Molfetta	GIBA [9]	X			TAFP
Cichorium intybus L. group *catalogna*	Cicoria rossa di Martina Franca	GIBA [9]	X			
Cichorium intybus L. group *catalogna*	Cicoria di Otranto	GIBA [9]	X			TAFP
Cichorium intybus L. group *catalogna*	Cicoria leccese	GIBA [9]	X			
Cucumis melo L.	Barattiere	GIBA [10]	X	X		TAFP
Cucumis melo L.	Carosello di Manduria	GIBA [10]	X			TAFP
Cucumis melo L.	Carosello mezzo lungo di Polignano	GIBA [10]	X	X		TAFP
Cucumis melo L.	"Cucummaru" de San Donato	X	X			TAFP
Cucumis melo L.	Meloncella bianca	GIBA [10]	X			TAFP
Cucumis melo L.	Meloncella fasciata	X	X			TAFP
Cucumis melo L.	Meloncella tonda di Galatina	X				TAFP
Cucumis melo L.	"Pilusella"	X	X			TAFP
Cucumis melo L.	"Poponeddha" di Corigliano	X	X			TAFP
Cucumis melo L.	"Spureddha"	X	X			TAFP
Cucumis melo L. subsp. *melo* group *flexuosus*	Tortarello	X				TAFP

Table 1. *Cont.*

Taxonomy	Local Variety Name	Characterization			Sanitation	Registration
		M	C	G		
Cucumis melo L. subsp. *melo* group *inodorus*	Melone bianco	GIBA [10]	X			TAFP
Cucumis melo L. subsp. *melo* group *inodorus*	Melone fior di fava	X				TAFP
Cucumis melo L. subsp. *melo* group *inodorus*	Melone "gaghiubbo"	X				TAFP
Cucumis melo L. subsp. *melo* group *inodorus*	Melone gialletto tondo estivo	GIBA [10]	X			TAFP
Cucumis melo L. subsp. *melo* group *inodorus*	Melone giallo allungato	GIBA [10]	X	X		TAFP
Cucumis melo L. subsp. *melo* group *inodorus*	Melone "minna de monaca"	GIBA [10]	X			TAFP
Cucumis melo L. subsp. *melo* group *inodorus*	Melone "scurzune"	GIBA [10]	X			TAFP
Cucumis melo L. subsp. *melo* group *inodorus*	Melone "zuccarinu"	GIBA [10]	X			TAFP
Cucumis sativus L.	Cetriolo mezzo lungo di Polignano	UPOV				TAFP
Cucurbita moschata Duch.	Cucuzza genovese	X				
Cucurbita pepo L.	Cocozza corritore bianca	GIBA [11]				
Cucurbita pepo L.	Cocozza corritore rigata	X				
Cucurbita pepo L.	Cocozza corritore striata	X				
Cynara cardunculus L. subsp. *scolymus* (L.) Hayek	Carciodo bianco di Ostuni	GIBA [12]		X		
Cynara cardunculus L. subsp. *scolymus* (L.) Hayek	Carciofo bianco di Taranto	GIBA [12]		X		
Cynara cardunculus L. subsp. *scolymus* (L.) Hayek	Carciofo brindisino	GIBA [12]		X	X	PGI
Cynara cardunculus L. subsp. *scolymus* (L.) Hayek	Carciofo centofoglie di Rutigliano	GIBA [12]		X		
Cynara cardunculus L. subsp. *scolymus* (L.) Hayek	Carciofo di Lucera	GIBA [12]	X	X	X	
Cynara cardunculus L. subsp. *scolymus* (L.) Hayek	Carciofo francesina	GIBA [12]		X	X	TAFP
Cynara cardunculus L. subsp. *scolymus* (L.) Hayek	Carciofo locale di Mola	GIBA [12]	X	X	X	TAFP
Cynara cardunculus L. subsp. *scolymus* (L.) Hayek	Carciofo nero a calice del Salento	GIBA [12]		X		
Cynara cardunculus L. subsp. *scolymus* (L.) Hayek	Carciofo nero di Ostuni	GIBA [12]		X	X	
Cynara cardunculus L. subsp. *scolymus* (L.) Hayek	Carciofo tricasino spinoso	GIBA [12]		X	X	
Cynara cardunculus L. subsp. *scolymus* (L.) Hayek	Carciofo verde di Putignano	GIBA [12]		X	X	TAFP
Cynara cardunculus L. subsp. *scolymus* (L.) Hayek	Carciofo violetto di Putignano	GIBA [12]		X	X	TAFP
Daucus carota L.	Carota di Margherita	X	X			
Daucus carota L.	Carota di Polignano	GIBA [13]	X			TAFP
Daucus carota L.	Carota di Tiggiano	GIBA [13]	X			TAFP
Eruca vesicaria (L.) Cav.	Rucola	X	X			
Foeniculum vulgare Mill. var. *azoricum*	Finocchio gigante di Bari	GIBA [14]				
Foeniculum vulgare Mill. var. *azoricum*	Finocchio nostrale barese	GIBA [14]				
Ipomoea batatas L.	Batata leccese	IBPGR	X			TAFP
Lactuca sativa L. group *capitata*	Lattuga romanella	GIBA [15]	X			
Phaseolus vulgaris L.	Fagiolino di Deliceto	X		X		
Pisum sativum L.	Pisello di Castellaneta	X				
Pisum sativum L.	Pisello di Monteiasi	X				
Pisum sativum L.	Pisello riccio di Sannicola	X				TAFP
Solanum lycopersicum L.	Pomodoro "a cancedd"	GIBA [16]	X			
Solanum lycopersicum L.	Pomodoro a foglia di patata	GIBA [16]	X			
Solanum lycopersicum L.	Pomodoro a mela di San Severo	GIBA [16]	X			
Solanum lycopersicum L.	Pomodoro "a pappacocu"	GIBA [16]	X			
Solanum lycopersicum L.	Pomodoro a perone	GIBA [16]	X			
Solanum lycopersicum L.	Pomodoro "a scresce"	X	X			
Solanum lycopersicum L.	Pomodoro a sole di Panni	GIBA [16]	X	X		
Solanum lycopersicum L.	Pomodoro barese	X				
Solanum lycopersicum L.	Pomodoro darseculo	IPGRI		X		
Solanum lycopersicum L.	Pomodoro della marina tipo piatto	GIBA [16]	X			
Solanum lycopersicum L.	Pomodoro della marina tipo tondo	GIBA [16]	X			
Solanum lycopersicum L.	Pomodoro di Crispiano	GIBA [16]	X			
Solanum lycopersicum L.	Pomodoro di Manduria	GIBA [16]	X	X		TAFP
Solanum lycopersicum L.	Pomodoro di Mola	GIBA [16]	X			TAFP
Solanum lycopersicum L.	Pomodoro di Morciano	GIBA [16]	X			TAFP
Solanum lycopersicum L.	Pomodoro di Panni allungato	X				
Solanum lycopersicum L.	Pomodoro di Torremaggiore	GIBA [16]	X			
Solanum lycopersicum L.	Pomodoro fiaschetto di Torre Guaceto	GIBA [16]	X	X		
Solanum lycopersicum L.	Pomodoro giallo d'inverno	GIBA [16]	X			TAFP
Solanum lycopersicum L.	Pomodoro giallo d'inverno di Torremaggiore	GIBA [16]	X			TAFP
Solanum lycopersicum L.	Pomodoro giallo da serbo di Monteleone e Panni	GIBA [16]	X			TAFP
Solanum lycopersicum L.	Pomodoro lamàsciano	GIBA [16]	X			

Table 1. *Cont.*

Taxonomy	Local Variety Name	Characterization			Sanitation	Registration
		M	C	G		
Solanum lycopersicum L.	Pomodoro leccese	GIBA [16]	X			
Solanum lycopersicum L.	Pomodoro "prunill"	GIBA[16]	X	X		
Solanum lycopersicum L.	Pomodoro racalino	GIBA [16]	X			
Solanum lycopersicum L.	Pomodoro regina	GIBA [16]	X	X		TAFP
Solanum lycopersicum L.	Pomodoro rosso d'inverno	GIBA [16]	X			TAFP
Solanum melongena L.	Melanzana antica	GIBA [17]				
Solanum melongena L.	Melanzana bianca	GIBA [17]				
Solanum melongena L.	Melanzana lunga di Vieste	X				
Solanum melongena L.	Melanzana marangiana di Zollino	X				
Solanum melongena L.	Melanzana nostrana di Palagianello	X				
Solanum tuberosum L.	Patata dei cortigli	X	X			
Vicia faba L.	Fava di Carpino	X	X			TAFP
Vicia faba L.	Fava di Zollino	X				TAFP
Vicia faba L.	Fava grande di Castellana Grotte	X				TAFP
Vicia faba L.	Fava "Nase 'n gule"	X				TAFP
Vicia faba L.	Fava nera di Monopoli	X				TAFP
Vicia faba L.	Fava romana	X				TAFP
Vigna unguiculata (L.) Walp. subsp. *unguiculata* (L.) Walp.	Fagiolino pinto	X	X	X		TAFP
Vigna unguiculata (L.) Walp. subsp. *unguiculata* (L.) Walp.	Fagiolino pinto a metro	X		X		
Vigna unguiculata (L.) Walp. subsp. *unguiculata* (L.) Walp.	Fagiolino pinto barese	X		X		TAFP
Vigna unguiculata (L.) Walp. subsp. *unguiculata* (L.) Walp.	Fagiolino pinto di Noci	X		X		TAFP
Vigna unguiculata (L.) Walp. subsp. *unguiculata* (L.) Walp.	Fagiolino pinto "mezza rama"	X		X		TAFP

M, morphological; C, chemical; G, genetic; GIBA, descriptors of the Italian Working Group for Biodiversity in Agriculture; TAFP, List of "Traditional Agri-Food Products" of the Italian Department for Agriculture; PGI, Protected Geographical Indication within European Community; UPOV, descriptors of the Union for the Protection of New Varieties of Plants; IBPGR, descriptors of the International Board for Plant Genetic Resources; IPGRI, descriptors of the International Plant Genetic Resources Institute; CPVO, Community Plant Variety Office; [1] Ref. CPVO TP/46/2; [2] Ref. CPVO TP/82/1; [3] Ref. UPOV TG/106/4; [4] Ref. UPOV TG/90/6; [5] Ref. CPVO TP/45/2; [6] Ref. CPVO TP/48/2; [7] Ref. CPVO TP/151/2; [8] Ref. CPVO TP/76/2; [9] Ref. UPOV TG/154/3; [10] Ref. CPVO TP/104/2; [11] Ref. CPVO TP/119/1; [12] Ref. CPVO TP/184/1; [13] Ref. CPVO TP/49/3; [14] Ref. CPVO TP/183/1; [15] Ref. CPVO TP/13/4; [16] Ref. CPVO TP/44/3; [17] Ref. CPVO TP/117/1.

This activity allowed also to obtain original results published in the international scientific literature and/or exploitable for future research (Table 2).

Table 2. Literature published targeting the different local varieties studied during project activities.

Local Variety	References
Barattiere	Pavan et al. [12]
tata leccese	Ruta and Lambardi [13]
Carciofo bianco di Taranto	Spanò et al. [14]
Carciofo brindisino	Curci et al. [15]
Carciofo cento foglie di Rutigliano	Ruta e Lambardi [12]
Carciofo di Lucera	Renna et al. [16]
Carciofo francesina	Spanò et al. [14]
Carciofo locale di Mola	Renna et al. [16]
	Spanò et al. [14]
Carciofo verde di Putignano	Spanò et al. [14]
Carciofo violetto di Putignano	Spanò et al. [14]
Carosello di Manduria	Pavan et al. [12]
	Cefola et al. [17,18]
Carota di Polignano	Renna et al. [19]
	Signore et al. [20]
Carota di Tiggiano	Scarano et al. [21]
Cicoria di Galatina	Renna et al. [22]
	Testone et al. [23]
Cicoria di Molfetta	Renna et al. [22]
	Testone et al. [23]
Melone giallo allungato	Girelli et al. [24]
Melone "minna de monaca"	Girelli et al. [24]
Melone "scurzune"	Girelli et al. [24]
Pomodoro di Manduria	Spanò et al. [25]

For example, Pavan et al. [12] used genotyping-by-sequencing (GBS) to characterize patterns of genetic diversity and genomic features within *Cucumis melo* L. Analyses of genetic structure, principal components, and hierarchical clustering supported the identification of three distinct subpopulations. One of them includes accessions of the local variety Carosello, referable to the *chate* taxonomic group. This is one of the oldest domesticated forms of *C. melo*, once widespread in Europe and now exposed to the risk of genetic erosion. The second subpopulation contains landraces of Barattiere, a regional vegetable production that had never been characterized at the DNA level and that was previously erroneously considered another form of *chate* melon. The third subpopulation includes genotypes of winter melon (*C. melo* var. *inodorus*). Genetic analysis within each subpopulation revealed patterns of diversity associated with fruit phenotype and geographical origin. In a study aimed to present the complete chloroplast genome sequence of the globe artichoke, obtained by a combination of data retrieved from genome and BAC clone sequencing, genomic DNA was extracted from young leaves of local variety Brindisino [15].

As regards qualitative and nutritional traits, compositional analysis and antioxidant properties were carried out for both Polignano and Tiggiano carrots, two colored landraces of *Daucus carota* L. The local varieties Molfettese and Galatina belong to the Catalogna chicory cultivated group (*Cichorium intybus* L.) and produce stems consumed at early growth stage as fresh or processed vegetables. These elongating inflorescence stems (so-called "*puntarelle*" that means small shoot tips) are consumed as fresh, fresh-cut (e.g., sliced into strips and bagged) or fully processed (e.g., cooked and glass jarred, or frozen) vegetables as ready-to eat foods [4,22]. In a context of request for product quality standards and uniformity, Molfettese and Galatina chicories were characterized for some nutritive and quality traits as both fresh and ready-to-use products [22]. These local varieties, used both raw and cooked, represent a refined and nutritious vegetables, because of the presence of several healthy compounds as well as their low nitrate content [22,23,26].

In Puglia some plant parts of vegetables, conventionally considered as by-products, are traditionally used as "unconventional" vegetables. For example, offshoots (so-called *cardoni* or *carducci*) of globe artichoke, produced during the vegetative growing cycle and removed by common cultural procedures (Figure 7), are consumed in the same way of cultivated cardoons (*Cynara cardunculus* L. var. *altilis* DC).

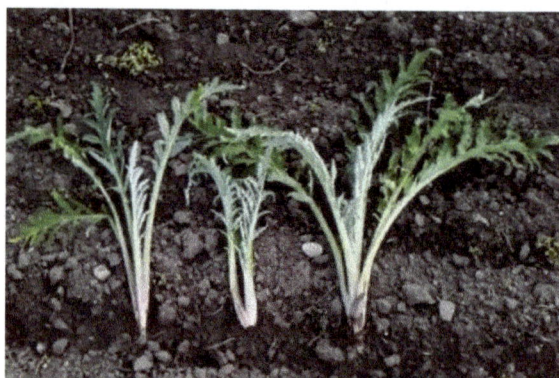

Figure 7. Offshoots (so-called *cardoni* or *carducci*) of a globe artichoke landrace.

The stems, petioles, flowers and smaller leaves of summer squash are used as greens (so-called *cime di zucchini*) (Figure 8), similar to other leafy vegetables such as chicory (*C. intybus* L.) and Swiss chard (*Beta vulgaris* L.).

Figure 8. Summer squash greens (so-called *cime di zucchini*).

In addition, the plant apex of faba bean, about 5–10 cm long, obtained from the green pruning, are used as greens (so-called *cime di fava*) (Figure 9) such as spinach leaves (*Spinacia oleracea* L.).

Figure 9. Faba greens (so-called *cime di fava*).

Renna et al. [16] showed that offshoots of globe artichoke, summer squash greens and faba greens have good potential as novel foods, being nutritious and refined products. In fact, for their content of fiber, offshoots of globe artichokes can be considered a useful food to bowel. Summer squash greens could be recommended as a vegetable to use especially in the case of hypoglycemic diets considering both content and composition of their carbohydrates. For their low content of nitrate, faba greens can be recommended as a substitute of nitrate-rich leafy vegetables.

A particular segment of Puglia's agro-biodiversity is represented by wild edible plants (WEP), which includes some progenitors of cultivated vegetables with which there is a continuum in the genetic profile [4]. WEP are a favorite delicacy in many countries and represent an extraordinary source of essential elements for the human health. They may be used to diversify and enrich modern diet with many colors and flavors, playing an important role in the diet of inhabitants in different parts of the world [27]. In Puglia, the harvesting of WEP is a time-honored custom and several species represent the essential ingredient to prepare some traditional food [28]. Renna et al. [27] determined thirteen elements (Na, K, Ca, Mg, Fe, Mn, Cu, Zn, Cr, Co, Cd, Ni and Pb) in 11 different WEP (*Amaranthus retroflexus, Foeniculum vulgare, Cichorium intybus, Glebionis coronaria, Sonchus* spp., *Borago officinalis, Diplotaxis tenuifolia, Sinapis arvensis, Papaver rhoeas, Plantago lagopus* and *Portulaca oleracea*) collected from countryside and urban areas of Bari. According to these authors, Renna [29] indicated that WEP may give a substantial contribution to the intake of mineral elements by consumers, representing a potentially good source in the daily diet.

Crenate broomrape (*Orobanche crenata* Forssk.) is a parasitic plant, which can cause serious damage to the production of legume crops in much of the Mediterranean basin. However, some authors [30] have reported its use as a food or folk drug, suggesting that it could be a refined food (Figure 10) with interesting nutritional traits. Thanks to WP 4 activity, Renna et al. [16] demonstrated that crenate broomrape may be considered as a very low source of dietary sodium as well as a vegetable with a low-middle content of nitrate. However, crenate broomrape showed a very interesting antioxidant activity, since a 100 g serving size supplies about 27 mMol of Trolox Equivalent [16].

Figure 10. Salad of crenate broomrape.

Sea fennel (*Crithmum maritimum* L.) is a perennial halophyte species typical of coastal ecosystems (Figure 11) used as a fresh ingredient for many food preparations [31]. It is listed as an item in the LTAFP of Puglia by the Italian Department for Agriculture.

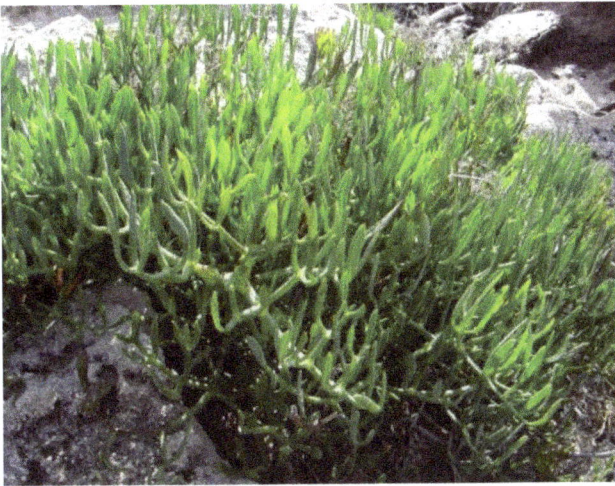

Figure 11. Plant of sea fennel on a rock-sandy beach as typical example of natural habitat for this halophyte.

Nevertheless, although considered as a promising biosaline crop, this halophyte is underutilized for commercial cultivation possibly due to a shortage of its consumer demand. Being also an aromatic herb, sea fennel plants may be used not only as fresh product but also as dried herbs. Thus, sea fennel has been characterized for some nutritive and quality traits as both fresh and dried product [32,33]. Based on these considerations, the possibility to cultivate the sea fennel is under consideration [34].

3.1.5. WP 5—Sanitation and Registration in the LTAFP

In Puglia, many globe artichoke ecotypes remained neglected and unnoticed for a long time and were progressively eroded by several causes, which include a poor phytosanitary status. Thanks to this activity a protocol for producing virus-free artichoke genetic resources for conservation, breeding, and production was carried out [14]. Eight local varieties of globe artichoke (Table 1) were sanitized from artichoke Italian latent virus (AILV), artichoke latent virus (ArLV) and tomato infectious chlorosis virus (TICV) by meristem-tip culture and in vitro thermotherapy through a limited number of subcultures to reduce the risk of "pastel variants" induction of and loss of earliness. A total of 25 virus-free primary sources were obtained and conserved ex situ in a nursery. It is expected that the use of these sanitized materials supplied by nursery-certified stocks and cultivated in open field for no more than two years should decrease inoculum potential and will ensure genetic resources for conservation, breeding and production [14].

As regards the registration of local varieties and related products in the List of "Traditional Agri-Food Products" (TAFP) of the Italian Department for Agriculture, 34 new registrations were carried out including fresh and processed vegetable products. In this regard, it is important to underline that, for all new registrations, it must be demonstrated that processing, preservation and ageing methods are consolidated in time and harmonious according to traditional rules, for a period of not less than 25 years. Thus, thanks to the accurate bibliographic and territorial investigation work of the WP 1, it was possible to acquire this essential information. Table 1 reports the state of art regarding Puglia local varieties with Protected Geographical Indication or registered as an item in the List of "Traditional Agri-Food Products" of the Italian Department for Agriculture.

3.1.6. WP 6—Database

Thanks to this activity, a computerized database called BiodiverSO Management System (BMS) was achieved. BMS allows to have a tool for correct and immediate management and processing of data of each project activity.

The dissemination of knowledge about (agro)biodiversity is a strategic factor in communicating the urgent need to defend and protect biological diversity. Thus, sharing and dissemination of the many collected information was promoted through the Project's WEB site (www.biodiversitapuglia.it), to which users access both to obtain information and report new local varieties (https://biodiversitapuglia.it/segnala-una-varietaspecie/). The Web GIS (https://biodiversitapuglia.it/webgis.php), with the cartographic collections, and the App for smartphones and tablets complete the usability of the information. Moreover, also social media, such as Facebook (https://www.facebook.com/BiodiverSO) and YouTube (https://www.youtube.com/user/BiodiversitaPuglia), were used for better sharing collected knowledge.

The interest in using informatic tools for collecting information on agro-biodiversity is increasing, since they greatly facilitate the collection and sharing of data. Anyway, in some cases, the use of such tools is not of immediate comprehension for a common user, for example, for the use of APIs, so an effort should be done to simplify these tools and allow greater participation for general public. In this optic, during this activity, we tested two tools freely available, namely Open Data Kit and Google Fusion Tables, to verify if their integrated use would allow a participative collection (and sharing) of data related to agro-biodiversity. Open Data Kit was used to collect information regarding several vegetable crops at risk of genetic erosion. This information, including multimedia and GPS data, were stored into the Google App Engine platform and afterwards was transferred into Google Fusion Tables for mapping and sharing with the general public. Both tools were tested in real scenarios in the Italian region of Puglia and the results seem to be encouraging: Open Data Kit has provided a good platform for the collection of data and it is reliable for georeferencing the fields, while Google Fusion Tables allowed us to show the data and share them in an easy way [35].

Within this activity, the use of Wikipedia was also proposed as a dissemination tool to show how to add/modify articles in Wikipedia for online divulgation and to demonstrate its validity by analyzing some data (pageviews, editing history, and the impact of Wikipedia as a referral toward the project's institutional website) related to the Wikipedia articles that were added/modified. Referrals from Wikipedia increased the visits to the institutional website by 30%, whereas the bounce rate decreased by 15%. Wikipedia may be a good tool to improve the dissemination of knowledge about (agro)biodiversity either online or offline, and the addition in Wikipedia's pages of scientific journal references and links to the projects' website may strengthen the diffusion of scientific knowledge [36].

3.1.7. WP 7—Data Sheets Writing

For all recovered local varieties, a descriptive sheet was realized and shared for users (https://biodiversitapuglia.it/varieta-orticole/). Moreover, with the aim to create a standardized system of identification, other data sheets (Figure 12) were realized by using descriptors from International Plant Genetic Resources Institute (IPGRI), International Union for the Protection of New Varieties of Plants (UPOV), International Board for Plant Genetic Resources (IBPGR) and Community Plant Variety Office (CPVO). Finally, genetic and nutritional data sheets were also realized by using results obtained from the WP 4 activity.

FONTI

http://www.upov.int/edocs/tgdocs/en/tg082.pdf

Figure 12. Examples of data sheets realized by using UPOV descriptors for a local variety of *Apium graveolens* L.

3.1.8. WP 8—*In situ* Conservation

As a result of this activity, 22 farms were funded for the in situ reproduction and cultivation. The financial support was provided for a five-year period to the so-called seed savers committed to in situ preserving the plant genetic resources. Moreover, some landraces and old open-pollinated varieties of vegetables have been kept alive within communities without any financial support by old farmers who act as seed savers (Figure 13).

Figure 13. An example of old farmer (Raffaele Cristoforo) who acts as a seed saver of the "Verza a cuore di bue", a local variety of cabbage (*Brassica oleracea* L. var. *capitata* [L.] DC.) (Photo credit: Beniamino Leoni)

3.2. Case Study

Until a few years ago, a multicolored landrace of *Daucus carota* L., known locally as yellow–purple Polignano carrot (PC) (Figure 14) was at high risk of genetic erosion. Therefore, under the BiodiverSO project, several tasks were performed within each of the eight project activities with the aim of verifying the effectiveness of these actions in terms of safeguarding this genetic resource strongly linked with local traditions.

Figure 14. Polignano carrots ready for market: it is interesting to note the great variety of the root colors.

Following bibliographic and territorial investigations (WP 1), documentary evidence supports that the PC has been cultivated for decades by local smallholder farmers in Southern Italy [17] around the town of Polignano a Mare, which is located in the southern part of the metropolitan area of Bari (41°0′0″ N 17°13′0″ E). The presence of carrots in Italy is established about the end of thirteenth or the beginning of the fourteenth century, probably from Arab countries through Venice, since it had an important trade with the Middle-East countries at that time [37]. Moreover, the presence of the carrot cultivation in Puglia is established at least since 1736 [18], while to the best of our knowledge it is possible to establish the cultivation of the PC at least since 1940 [17]. Thanks to this activity, several knowledge and traditions related to the crop cycle of PC were retrieved and shared using social media (https://www.youtube.com/watch?v=S_fYKNnkkE4). The "seed" (it is an achene actually) used for planting is produced every year by the smallholder farmers. In late summer, from mid-August to mid-September, sowing is performed by a mechanical seeder placing each seed at a depth of 2–3 cm, about 2–3 cm apart with a distance of 30 cm between rows, resulting in a seeding density of 90–120 plants m^{-2} [19]. Soil fertility remaining from the previous crop is sufficient to satisfy the needs of the PC, so the PC requires no fertilization. Because of the proximity of the fields to the coast, irrigation water, extracted from underground wells (at a depth of 15–20 m), is typically brackish with electrical conductivity (EC) ranging from 3 to 4 dS m^{-1} [17]. The relatively high EC of the irrigation water is probably the main factor positively affecting carrot flavor. Harvesting normally starts in mid-December and lasts until April–May, and is performed manually using a pitchfork and is probably the most difficult step in the cultivation of this landrace [17].

Seeds of PC (Figure 15) were recovered (WP 2) to be used for ex situ conservation (WP 3) through long-time seeds preservation in seed banks as well as by in vitro techniques (Figure 16). In addition, a catalog field was realized by using soilless systems within the greenhouse (Figure 17) of the "La Noria" experimental farm - National Research Council, Bari, Italy.

Figure 15. Seeds of Polignano carrot with indication of size (Photo credit: Beniamino Leoni).

Figure 16. Seedlings of Polignano carrot obtained by micropropagation (Photo credit: Claudia Ruta—from https://biodiversitapuglia.it).

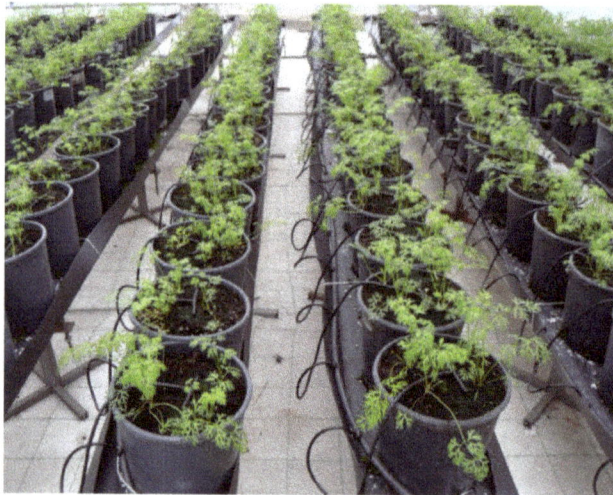

Figure 17. Catalog field of Polignano carrot realized by using soilless systems within the greenhouse of the "La Noria" experimental farm—National Research Council, Bari, Italy.

As regards characterization (WP 4), PC showed a total glucose, fructose and sucrose content (on average 4.38 g/100 g FW) about 22% lower than orange commercial varieties [17]. On the other hand, this landrace showed a relative sweetness similar to some commercial carrots [17], because the highest contributor to the relative sweetness in the PC is fructose (43.5%) which contributes to its distinctive flavor as well as to its glycemic index [17]. The roots of the PC have various skin colors which allow them to be separated into three groups: yellow, orange and purple. The purple form of PC showed an antioxidant activity (on average 42.7 mg of Trolox equivalent/100 g FW) about tenfold higher than yellow and orange types [17]. Likewise, the total amount of phenolics in the purple PC (67.6 mg/100 g FW, on average) was fourfold higher than in yellow and orange types [17].

Regarding the content of total carotenoid, the purple type of PC showed an amount (on average 43.3 mg/100 g FW) 5.6- and 4.7-fold higher than yellow and orange types, respectively, while the β-carotene content in the purple type (on average 15.5 mg/100 g FW) was 3.5- and 5.9-fold higher compared to yellow and orange PC, respectively. Following physiological analysis, a drawback showed by this landrace is the short shelf-life of the roots, even at low temperatures, because of its high respiration rate; its roots are therefore considered perishable and difficult to store as a fresh product [38].

Thanks to the information obtained by WP 1 activity, in 2015, the PC was registered (WP 5) in the LTAFP of the Italian Department for Agriculture since it was demonstrated that its cultivation and processing are consolidated in time and harmonious according to traditional rules at least from 25 years.

Sharing and dissemination of the many collected information on database regarding PC (WP 6) was promoted also through Web 2.0 tools. A Wikipedia page about the PC was created from scratch (http://wiki.biodiversitapuglia.it/Carota_di_Polignano), receiving more than 15,000 page views in less than two years [25]. Moreover, a descriptive sheet was also realized and shared for users (https://biodiversitapuglia.it/varieta-orticole/carota-di-polignano/).

Finally, the in situ conservation (WP 8) is realized by pensioners (60–75 years old) who act as seed savers (Figure 18) on very small areas. For this reason, no farmer has applied for agri-environmental funding under Puglia RDP. Such an omission puts this ancient local variety at potential higher risk of genetic erosion in comparison with other genetic resources. To face this risk, smallholder farmers consumers and processors of this landrace have created the association "La bastinaca di San Vito", and have added it to the Slow Food Presidia of traditional products. In this regard, the Presidia seek to promote sustainable farming systems and, in particular, the protection of the local varieties at risk of genetic erosion as well as local traditions. The Slow Food Presidia is a project that aims to protect a community of farmers and promote their artisan food products. In this way, it is possible to stimulate market opportunities and valorize a territory through the use of a model of sustainable and environmentally friendly agriculture. At the same time, local traditions and cultural identity of people are also preserved and promoted [39].

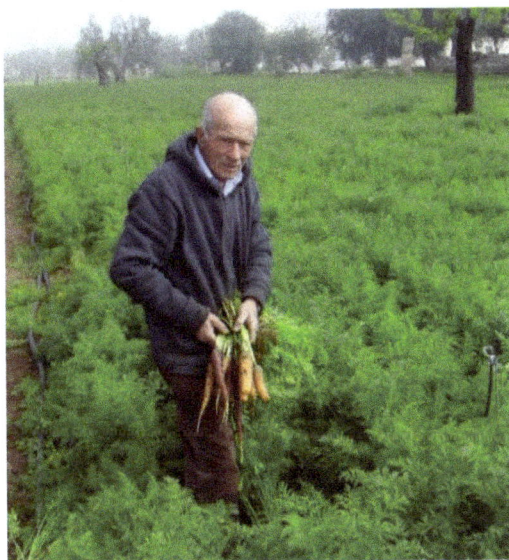

Figure 18. A local old seed savers (Oronzo Giuliacci) of the Polignano carrot.

4. Conclusions

The Puglia RDP is an example of protection and recovery of vegetables at risk of genetic erosion, and its implementation through integrated projects such as BiodiverSO could help to identify and valorize much of these plant germplasms. Nevertheless, the financial support to seed savers for realizing the in situ conservation is too limited if aimed to provide an adequate incentive for cultivation. Therefore, it is important to highlight that in situ conservation of genetic resources needs to be based not only on institutional programs, but mainly on the possibility, especially for young growers, to use these resources for productive activities which would imply a real income. Moreover, the potential strategy of integrating institutional programs with research and development projects, could be a model useful to preserve the biodiversity of vegetable crops.

Author Contributions: Conceptualization, M.R. and P.S.; Methodology, M.R. and P.S.; Investigation, M.R., A.S., M.G. and P.S.; Writing—Original Draft Preparation, M.R., F.M. and P.S.; Writing—Review and Editing, M.R., F.M., A.S., M.G. and P.S.; Supervision, M.R. and P.S.; Project Administration, P.S.; and Funding Acquisition, P.S.

Funding: This research was funded by Regione Puglia Administration under Rural Development Program 2014–2020—Project "Biodiversity of vegetable crops in Puglia (BiodiverSO)", Measure 10, Sub measure 10.2, Operation 1 "Program for the conservation and the valorization of the genetic resources in agriculture".

Acknowledgments: The authors thank Beniamino Leoni for Figures 3, 5, 13 and 15 and Claudia Ruta for Figures 6 and 16.

Conflicts of Interest: The authors declare no conflict of interest.

References

1. Attitudes of Europeans towards Biodiversity. Available online: http://ec.europa.eu/COMMFrontOffice/publicopinion/index.cfm/Survey/index#p=1&search=biodiversity (accessed on 4 June 2018).
2. FAO. What is agrobiodiversity? In *Building on Gender, Agrobiodiversity and Local Knowledge*; FAO: Rome, Italy, 2004; pp. 1–18.
3. Silva, J. *Biodiversity and vegetable breeding in the light of developments in intellectual property rights. Ecosystems Biodiversity*; In Tech: Rijeka, Croatia, 2011; pp. 389–428.
4. Elia, A.; Santamaria, P. Biodiversity in vegetable crops, a heritage to save: The case of Puglia region. *Ital. J. Agron.* **2013**, *8*, 4. [CrossRef]
5. Zeven, A.C. Landraces: A review of definitions and classifications. *Euphytica* **1998**, *104*, 127–139. [CrossRef]
6. ISTAT. Electronic Information System on Agriculture and Livestock. Available online: http://agri.istat.it/ (accessed on 4 June 2018).
7. Bio in Cifre. Available online: http://www.sinab.it/sites/default/files/share/OK%21%21.pdf (accessed on 4 June 2018).
8. Hammer, K.; Laghetti, G. Genetic erosion—Examples from Italy1,2. *Genet. Resour. Crop Evol.* **2005**, *52*, 629–634. [CrossRef]
9. Laghetti, G.; Fiorentino, G.; Hammer, K.; Pignone, D. On the trail of the last autochthonous Italian einkorn (*Triticum monococcum* L.) and emmer (*Triticum dicoccon* Schrank) populations: a mission impossible? *Genet. Resour. Crop Evol.* **2009**, *56*, 1163–1170. [CrossRef]
10. Egidi, P. *La Colonia Saracena Di Lucera E La Sua Distruzione*; PopCorn Press: Lucera, Italy, 1915.
11. Accogli, R.; Conversa, G.; Ricciardi, L.; Sonnante, G.; Santamaria, P. *Nuovo Almanacco BiodiverSO*; Università degli Studi di Bari Aldo Moro: Bari, Italy, 2018; ISBN 978-88-6629-024-7.
12. Pavan, S.; Marcotrigiano, A.R.; Ciani, E.; Mazzeo, R.; Zonno, V.; Ruggieri, V.; Lotti, C.; Ricciardi, L. Genotyping-by-sequencing of a melon (*Cucumis melo* L.) germplasm collection from a secondary center of diversity highlights patterns of genetic variation and genomic features of different gene pools. *BMC Genom.* **2017**, *18*, 59. [CrossRef] [PubMed]
13. Ruta, C.; Lambardi, M. La coltura in vitro per la conservazione della biodiversità orticola. *Italus Hortus* **2018**, *25*, 53–74.
14. Spanò, R.; Bottalico, G.; Corrado, A.; Campanale, A.; Di Franco, A.; Mascia, T. A protocol for producing virus-free artichoke genetic resources for conservation, breeding, and production. *Agriculture* **2018**, *8*, 36. [CrossRef]

15. Curci, P.L.; De Paola, D.; Danzi, D.; Vendramin, G.G.; Sonnante, G. Complete chloroplast genome of the multifunctional crop globe artichoke and comparison with other asteraceae. *PLoS ONE* **2015**, *10*, e0120589. [CrossRef] [PubMed]

16. Renna, M.; Signore, A.; Paradiso, V.M.; Santamaria, P. Faba greens, globe artichoke's offshoots, crenate broomrape and summer squash greens: unconventional vegetables of puglia (southern italy) with good quality traits. *Front. Plant Sci.* **2018**, *9*, 1–13. [CrossRef] [PubMed]

17. Cefola, M.; Pace, B.; Renna, M.; Santamaria, P.; Signore, A.; Serio, F. Compositional analysis and antioxidant profile of yellow, orange and purple Polignano carrots. *Ital. J. Food Sci.* **2012**, *24*, 284–291.

18. Cefola, M.; Mariani, R.; Pace, B.; Renna, M.; Santamaria, P.; Serio, F.; Signore, A. La carota di Polignano. In *La biodiversità delle colture pugliesi*; INEA: Bari, Italy, 2013; pp. 60–73.

19. Renna, M.; Serio, F.; Signore, A.; Santamaria, P. The yellow–purple Polignano carrot (*Daucus carota* L.): A multicoloured landrace from the Puglia region (Southern Italy) at risk of genetic erosion. *Genet. Resour. Crop Evol.* **2014**, *61*, 1611–1619. [CrossRef]

20. Signore, A.; Renna, M.; D'Imperio, M.; Serio, F.; Santamaria, P. Preliminary evidences of biofortification with iodine of "carota di polignano," an italian carrot landrace. *Front. Plant Sci.* **2018**, *9*, 1–8. [CrossRef] [PubMed]

21. Scarano, A.; Gerardi, C.; D'Amico, L.; Accogli, R.; Santino, A. Phytochemical analysis and antioxidant properties in colored tiggiano carrots. *Agriculture* **2018**, *8*, 102. [CrossRef]

22. Renna, M.; Gonnella, M.; Giannino, D.; Santamaria, P. Quality evaluation of cook-chilled chicory stems (*Cichorium intybus* L., Catalogna group) by conventional and *sous vide* cooking methods. *J. Sci. Food Agric.* **2014**, *94*, 656–665. [CrossRef] [PubMed]

23. Testone, G.; Mele, G.; Di Giacomo, E.; Gonnella, M.; Renna, M.; Tenore, G.C.; Nicolodi, C.; Frugis, G.; Iannelli, M.A.; Arnesi, G.; et al. Insights into the sesquiterpenoid pathway by metabolic profiling and de novo transcriptome assembly of stem-chicory (*Cichorium intybus* Cultigroup "Catalogna"). *Front. Plant Sci.* **2016**, *7*, 1676. [CrossRef] [PubMed]

24. Girelli, C.R.; Accogli, R.; Del Coco, L.; Angilè, F.; De Bellis, L.; Fanizzi, F.P. ^{1}H-NMR-based metabolomic profiles of different sweet melon (*Cucumis melo* L.) Salento varieties: Analysis and comparison. *Food Res. Int.* **2018**, *114*, 81–89. [CrossRef]

25. Spanò, R.; Gallitelli, D.; Mascia, T. Grafting to manage infections of top stunting and necrogenic strains of cucumber mosaic virus in tomato. *Ann. Appl. Biol.* **2017**, *171*, 393–404. [CrossRef]

26. D'Acunzo, F.; Giannino, D.; Longo, V.; Ciardi, M.; Testone, G.; Mele, G.; Nicolodi, C.; Gonnella, M.; Renna, M.; Arnesi, G.; Schiappa, A.; Ursini, O. Influence of cultivation sites on sterol, nitrate, total phenolic contents and antioxidant activity in endive and stem chicory edible products. *Int. J. Food Sci. Nutr.* **2017**, *68*, 52–64. [CrossRef] [PubMed]

27. Renna, M.; Cocozza, C.; Gonnella, M.; Abdelrahman, H.; Santamaria, P. Elemental characterization of wild edible plants from countryside and urban areas. *Food Chem.* **2015**, *177*, 29–36. [CrossRef] [PubMed]

28. Renna, M.; Rinaldi, V.A.; Gonnella, M. The Mediterranean Diet between traditional foods and human health: The culinary example of Puglia (Southern Italy). *Int. J. Gastron. Food Sci.* **2015**, *2*, 63–71. [CrossRef]

29. Renna, M. Wild edible plants as a source of mineral elements in the daily diet. *Prog. Nutr.* **2017**, *19*, 219–222.

30. Renna, M.; Serio, F.; Santamaria, P. Crenate broomrape (*Orobanche crenata* Forskal): prospects as a food product for human nutrition. *Genet. Resour. Crop Evol.* **2015**, *62*, 795–802. [CrossRef]

31. Renna, M.; Gonnella, M. The use of the sea fennel as a new spice-colorant in culinary preparations. *Int. J. Gastron. Food Sci.* **2012**, *1*, 111–115. [CrossRef]

32. Renna, M.; Gonnella, M.; Caretto, S.; Mita, G.; Serio, F. Sea fennel (*Crithmum maritimum* L.): from underutilized crop to new dried product for food use. *Genet. Resour. Crop Evol.* **2017**, *64*, 205–216. [CrossRef]

33. Giungato, P.; Renna, M.; Rana, R.; Licen, S.; Barbieri, P. Characterization of dried and freeze-dried sea fennel (*Crithmum maritimum* L.) samples with headspace gas-chromatography/mass spectrometry and evaluation of an electronic nose discrimination potential. *Food Res. Int.* **2018**, in press. [CrossRef]

34. Montesano, F.F.; Gattullo, C.E.; Parente, A.; Terzano, R.; Renna, M. Cultivation of potted sea fennel, an emerging mediterranean halophyte, using a renewable seaweed-based material as a peat substitute. *Agriculture* **2018**, *8*, 96. [CrossRef]

35. Signore, A. Mapping and sharing agro-biodiversity using Open Data Kit and Google Fusion Tables. *Comput. Electron. Agric.* **2016**, *127*, 87–91. [CrossRef]

36. Signore, A.; Serio, F.; Santamaria, P. Wikipedia as a tool for disseminating knowledge of (agro)biodiversity. *Horttechnology* **2014**, *24*, 118–126.

37. Banga, O. The development of the original European carrot material. *Euphytica* **1957**, *6*, 64–76. [CrossRef]

38. Renna, M.; Pace, B.; Cefola, M.; Santamaria, P.; Serio, F.; Gonnella, M. Comparison of two jam making methods to preserve the quality of colored carrots. *LWT-Food Sci. Technol.* **2013**, *53*, 547–554. [CrossRef]

39. Renna, M. From the farm to the plate: Agro-biodiversity valorization as a tool for promoting a sustainable diet. *Prog. Nutr.* **2015**, *17*, 77–80.

agriculture

MDPI

Article

Patterns of Genetic Diversity and Implications for In Situ Conservation of Wild Celery (*Apium graveolens* L. ssp. *graveolens*)

Lothar Frese *, Maria Bönisch, Marion Nachtigall and Uta Schirmak

Julius Kühn-Institut (JKI), Federal Research Centre for Cultivated Plants, Institute for Breeding Research on Agricultural Crops, Erwin-Baur-Str. 27, 06484 Quedlinburg, Germany; maria.boenisch@julius-kuehn.de (M.B.); marion.nachtigall@julius-kuehn.de (M.N.); uta.schirmak@julius-kuehn.de (U.S.)
* Correspondence: lothar.frese@julius-kuehn.de; Tel.: +49-3946-47701

Received: 26 May 2018; Accepted: 16 August 2018; Published: 22 August 2018

Abstract: In Germany, the wild ancestor (*Apium graveolens* L. ssp. *graveolens*) of celery and celeriac is threatened by genetic erosion. Seventy-eight potentially suitable genetic reserve sites representing differing ecogeographic units were assessed with regard to the conservation status of the populations. At 27 of the 78 sites, 30 individual plants were sampled and genetically analyzed with 16 polymorphic microsatellite makers. The Discriminant Analysis of Principal Components (DAPC) was applied to identify clusters of genetically similar individuals. In most cases (25 out of 27 occurrences) individuals clustered into groups according to their sampling site. Next to three clearly separated occurrences (AgG, AgUW, AgFEH) two large groups of inland and Baltic Sea coast occurrences, respectively, were recognized. Occurrences from the coastal part of the distribution area were interspersed into the group of inland occurrences and vice versa. The genetic distribution pattern is therefore complex. The complementary compositional genetic differentiation Δ_j was calculated to identify the Most Appropriate Wild Populations (MAWP) for the establishment of genetic reserves. Altogether 15 sites are recommended to form a genetic reserve network. This organisational structure appears suitable for promoting the in situ conservation of intraspecific genetic diversity and the species' adaptability. As seed samples of each MAWP will be stored in a genebank, the network would likewise contribute to the long-term ex situ conservation of genetic resources for plant breeding.

Keywords: *Apium graveolens*; genetic resources; crop wild relative; in situ conservation; microsatellite marker; genetic distance; genetic differentiation; genetic reserve

1. Introduction

Celery (*Apium graveolens* L. var. *dulce*) and celeriac (*Apium graveolens* L. var. *rapaceum*) have significant global economic importance. The total production value of celery amounted 458 million US$ in the USA in the year 2013 [1]. In the European Union (EU-28) celery/celeriac was grown on an average of 8125 ha within the period from 2011–2016 [2]. Due to cultural preferences, different crop types are grown and consumed. *A. graveolens* var. *rapaceum* forms aromatic tubers which can be stored as a winter vegetable and which is grown in the colder regions of Central and Eastern Europe. Consumers in Western Europe, the Mediterranean region, India, China, and the USA prefer the petiole celery (var. *dulce*). A further variety (var. *secalinum*) is used as a spice. Processed dried leaves or seeds are an important component of condiments [3,4].

All modern celery varieties have been derived from only two forms ('White Plume' and 'Giant Pascal') [5]. According to Melchinger and Lübberstedt [6] and Domblides et al. (cit. in [7]) the genetic base of celery breeding is small and may therefore impede further breeding progress. Wang et al. [8] developed sequence related amplified polymorphism (SRAP) markers and used them along with microsatellite markers (SSR) published by Acquadro et al. [9] to investigate the genetic differences between European and Chinese celery varieties (var. *dulce*). They detected a unique genetic signature within the Chinese germplasm [8]. Using transcriptome sequences Fu et al. [10,11] developed a large set of EST-SSR to study genetic diversity within the cultivated species.

European genebanks conserve 830 accessions of cultivated *Apium* germplasm [12]. A large number of ex situ accessions but also a limited number of landraces maintained by farmers in traditional seed supply systems [7] is available for base-broadening. Spoor and Simmonds discern between two base-broadening efforts: the incorporation approach aims at increasing the genetic variability of the genetic background while the introgression approach involves the transfer of specific genes from a donor into the breeding material by crossing and backcrossing [13].

Introgression programmes commonly target distinct quality and resistance traits. Because celery consumption can provoke severe allergic reactions, foods and condiments containing celery must be clearly marked as such. Non-specific lipid transfer proteins (nsLTP) designated Api g 1 [14], Api g 2 [15] and Api g 6 [16] trigger the allergic reaction. The development of celery varieties which do not cause allergic reaction would be an important objective of quality breeding. However, non-allergic variants of cultivated forms or wild species have not yet been discovered.

As for most crops, breeding of pest- and disease-resistant celery varieties is an important aim for two reasons. Global trade and climate change promote the spread of new pests and diseases. Additionally, at the same time the acceptance for chemical plant protection is decreasing. Therefore, breeding of resistant varieties becomes increasingly important. Ochoa and Quiros used *A. panul* Reiche and *A. chilense* Hook. and Arn. to improve the resistance of the crop to the leaf spot disease (*Septoria apicola* Speg. f.s. *apii*) [17]. *A. prostratum* Labill. ex Vent is resistant to *Spodoptera exigua* Hübner, the beet army worm the larvae of which damage the leaves [18]. *Helosciadium nodiflorum* (L.) W. D. J. Koch (Syn. *A. nodiflorum* L.) is characterized by a resistance inhibiting the development of the leaf miner *Lyriomiza trifolii* Burgess beyond the larvae stage [19,20].

For *A. graveolens* ssp. *graveolens* (2n = 22) [21], the close wild relative of celery/celeriac, information on traits of interest to breeders is not available. European genebanks conserve only 51 accessions of this species (sample status wild) of which 59% were collected and are conserved in Portugal. The remaining 41% of the European holding is shared by genebanks in Azerbaijan, Czech Republic, Germany, Israel, Spain and the United Kingdom [12].

The species is native to most of Europe, North Africa, Siberia and the Caucasus and has been reported by a total of 56 countries to be located in these regions [22]. Obviously, the current holding does not at all cover the full range of ecogeographic variation of the distribution area. Further collecting of wild *A. graveolens* is required to capture a representative sample of genetic diversity of the species. A start was made in Germany in the context of a project aiming at the establishment of genetic reserves for four native wild celery species, namely *A. graveolens* L. ssp. *graveolens*, *Helosciadium repens* (Jacq.) W. D. J. Koch (Syn. *A. repens* (Jacq.) Lag.), *H. nodiflorum* (Syn. *A. nodiflorum* L.) and *H. inundatum* (L.) W. D. J. Koch (Syn. *A. inundatum* (L.) Rchb.f.) [23].

A. graveolens ssp. *graveolens* mainly occurs in northern and central Germany. In the northern part *A. graveolens* can be found in the hinterland of the coast on salt meadows and saline pasture land as well as in brackish water stands of *Phragmites australis* [24,25]. In the central part the species grows around natural salt water springs, along saline creeks or on wet sites at the foot of salty spoil heaps [26–28].

The ultimate goal of the project consists in the establishment of a network of persons and institutions managing a set of genetic reserve sites in Germany. A European group of researchers is pursuing similar interests within the framework of the Farmer's Pride project [29]. We expect that the national and European activity will contribute to the development of an integrated European crop wild relative conservation strategy as outlined in the concept paper by Maxted et al. [30].

A genetic reserve is defined as "The location, management and monitoring of genetic diversity in natural populations within defined areas designated for long-term active conservation" [31]. It is a conservation procedure combining the dynamic conservation of a population in the habitat (in situ) with the static preservation of a sample taken from the population in a genebank (ex situ). According to the proposed quality standards for genetic reserves, Most Appropriate (crop) Wild (relative) Populations (MAWP) [30] have to be identified through a rigorous scientific process which includes the genetic characterization of populations of the target taxon [32]. The genetic analysis is required to determine those populations representing altogether the genetic diversity of the target taxon best. Populations need to fulfil additional quality criteria in order to be proposed as MAWP. The population should exist at the location for at least ten years (native or introduced), contain specific traits of interest, be managed according to the minimum quality standards for genetic reserve conservation proposed by Iriondo et al. [32] and nominated as MAWP by a responsible national agency [30].

Kell et al. suggested a stepwise procedure to identify suitable genetic reserve sites [33]. The stepwise decision-making process ends with the designation of genetic reserve sites for a set of MAWPs and the creation of MAWP accessions for ex situ conservation and use. The objectives of our study were (i) to analyze the genetic structures within a collection of *A. graveolens* ssp. *graveolens*, (ii) to identify MAWPs based on the representativity or uniqueness of the genetic composition of occurrences and (iii) to locate sites within the German distribution area suited for the establishment of genetic reserves.

Generally, species are subdivided into plant groups more or less strongly connected by gene flow. Declining gene flow with increasing spatial distance between occurrences combined with adaptation to specific local environmental conditions can result in genetic differentiation between occurrences. In addition, genetic drift caused by local extinction and re-establishment of occurrences shapes their genetic composition [34]. Molecular genetic tools have been extensively used to study patterns of genetic diversity in many species [35]. The development of molecular markers for celery started in 1984 [36]. Huestis et al. [37] developed 21 RFLP markers and mapped them to eight linkage groups. Until recently, the development of a greater number of genetic markers for a crop of limited global economic importance based on whole-genome sequences seemed infeasible due to the large genome size of 3×10^9 bp [10]. This assumption was outdated quickly. Rapid technological progress allowed sequencing the whole genome of 'Q2-JN11' celery, a highly inbred line of 'Jinnan Shiqin'. The online database of the whole-genome sequences, CeleryDB, was constructed and the sequence information made available to the research community [38]. Sequences of unigenes described by Li et al. [39] are available in the celery database and can serve as a reference base to identify trait-associated single nucleotide polymorphism and additional SSR markers. The technological advancements allowed the authors of Reference [10] to analyze a large set of transcriptome sequences and to develop a set of EST-SSRs for breeding research and population genetics. The latter deals with the analysis of the distribution of genetic variation within and between populations.

Our study was initiated to support in situ conservation actions. To this end we determined the structures of genetic diversity within the sampled material as well as the differences in trait distribution between occurrences. The latter measurement of genetic variation is termed compositional differentiation and is of particular relevance because it allows the identification of MAWP candidates [40,41].

2. Results

In Germany, *A. graveolens* is mainly distributed in the coastal region of the Baltic Sea and in the centre of Germany where the species occurs in haline marshlands. The genetic differences between 27 occurrences sampled in the distribution area (Table 1) were analyzed with 16 SSR markers to identify candidate MAWPs and sites suited for the establishment of genetic reserves. The descriptive marker parameters are presented in Table 2. The number of distinct alleles ranges between 4 (marker Fn09) and 18 (marker ECMS6, QC43), and the polymorphic information content (PIC) ranges between 0.042 (marker Fn100) and 0.803 (marker ECMS39). Mostly, high PIC values indicate appropriateness of respective markers for genetic analyses. The observed heterozygosity (H_o) ranges between 0.005 (Fn100) and 0.963 (QC75) and the expected (H_e) heterozygosity between 0.042 (Fn100) and 0.826 (ECMS39).

Table 1. Geographic origin of 27 occurrences of *Apium graveolens* sampled in Germany. The laboratory and population identifier as well as the number of plants analyzed with 16 SSR markers is presented. The term ecogeographic unit relates to the German term "Naturraum 3. Grades" [42]. The laboratory identifiers in bold letter type mark Most Appropriate Wild Populations (MAWP). Arguments for and against the MAWP nomination are given in the last column.

LabID	Population ID	Location	Ecogeographic Unit	No. of Analyzed Individuals	Arguments For or Against the Nomination as Candidate MAWPs
AgBW	BW-UB-20150728-0934	Baden-Württemberg; Ubstadt-Weiher (samples from ex situ cultivation in Botanical Garden of KIT and Tübingen)	Gäuplatten im Neckar- und Tauberland	30	The only representative of the ecogeographic unit, most southern occurrence in Germany, site management exists already
AgSEHL	SH-SEHL-20150824-0900, -1000	Schleswig-Holstein: Sehlendorf	Schleswig-Holsteinisches Hügelland	30	Adjustment of management would be required
AgEICH	SH-EICH-20150824-1000, -1010, -1100, -1110, -1120, -1200	Schleswig-Holstein: Eichholzniederung	Schleswig-Holsteinisches Hügelland	30	Large site and population size, site management exists already
AgKREM	SH-KREM-20150902-1315, -1500, -1510, -1520	Schleswig-Holstein: Neustadt	Schleswig-Holsteinisches Hügelland	30	Within the same ecogeographic unit AgEICH matches genetic reserves quality criteria better than AgKREM
AgFEH	SH-FEH-20150902-1000, -1010, -1020	Schleswig-Holstein: Fehmarn	Schleswig-Holsteinisches Hügelland	30	Within the same ecogeographic unit AgEICH matches genetic reserves quality criteria better than AgFEH
AgOEH	SH-OEH-20150826-1200, 1300, -1320, -1330, -1340, -1350, -1400	Schleswig-Holstein: Schleimündung	Schleswig-Holsteinisches Hügelland	30	Within the same ecogeographic unit AgEICH matches genetic reserves quality criteria better than AgOEH
AgBBG	Bbg-JÜ-20150727-1739	Brandenburg: Gröben	Mittelbrandenburgische Platten und Niederungen	30	The only representative of the ecogeographic unit, site management exists already
AgFRI	ST-FRIED-20150621-1437	Sachsen-Anhalt: Friedeburg	Mitteldeutsches Schwarzerdegebiet	30	Within the same ecogeographic unit AgHEC matches genetic reserves quality criteria better than AgFRI
AgWA	ST-WALBE-20150720-1304	Sachsen-Anhalt: Walbeck	Weser-Aller-Flachland	30	The only representative of the natural region, highest deviation in genetic composition from the complement
AgHEC	ST-HECBE-20150709-0957, ST-HECSA-20150709-0854	Sachsen-Anhalt: Hecklingen	Mitteldeutsches Schwarzerdegebiet	30	Represents the genetic composition of the complement best, site management exists already
AgZIE	ST-ZIELI-20150709-1223	Sachsen-Anhalt: Zielitz	Elbalniederung	30	The only representative of the ecogeographic unit
AgSUE	ST-SÜLLW-20150714-1054, ST-SÜLNO-20150714-1003	Sachsen-Anhalt: Sülldorf	Mitteldeutsches Schwarzerdegebiet	30	Large site and population size, site management exists already

Table 1. *Cont.*

LabID	Population ID	Location	Ecogeographic Unit	No. of Analyzed Individuals	Arguments For or Against the Nomination as Candidate MAWPs
AgROS	ST-ROSSL-20150702-1029	Thüringen/Sachsen-Anhalt: Roßleben	Thüringer Becken und Randplatten	30	The only representative of the ecogeographic unit
AgBEN	ST-BENWÜ-20150702-0828	Sachsen-Anhalt: Bennstedt	Mitteldeutsches Schwarzerdegebiet	30	Within the same ecogeographic unit AgHEC matches genetic reserves quality criteria better than AgBEN
AgWU	MV-WU-20150824-1100	Mecklenburg-Vorpommern: Halbinsel Wustrow	Mecklenburgisch-Vorpommmersches Küstengebiet	30	Large site and population, best representative of the ecogeographic unit
AgDA	MV-DA-20150901-0900	Mecklenburg-Vorpommern: Dabitz	Mecklenburgisch-Vorpommmersches Küstengebiet	23	Critically small population size. Within the same ecogeographic unit AgDZ matches genetic reserves quality criteria better than AgDA
AgDZ	MV-DSZ-20150903-1430	Mecklenburg-Vorpommern: Dassow	Mecklenburgische Seenplatte	30	Best representative of the ecogeographic unit
AgUW	MV-UW-20150902-1000	Mecklenburg-Vorpommern: Rostock	Mecklenburgisch-Vorpommmersches Küstengebiet	30	Within the same ecogeographic unit AgDZ matches genetic reserves quality criteria better than AgUW
AgHEU	MV-HEU-20150921-1600	Mecklenburg-Vorpommern: Bodden bei Rügen	Mecklenburgisch-Vorpommmersches Küstengebiet	30	Access to population is complicated. Within the same ecogeographic unit AgDZ matches genetic reserves quality criteria better than AgHEU
AgG	MV-G-20150831-1200	Mecklenburg-Vorpommern: Gristow	Nordostmecklenburgisches Flachland	30	The only representative of the ecogeographic unit
AgHID	MV-HiABs-20150912-1400	Mecklenburg-Vorpommern: Hiddensee	Mecklenburgisch-Vorpommmersches Küstengebiet	30	Within the same ecogeographic unit AgDZ matches genetic reserves quality criteria better than AgHID
AgSL	MV-Su-20150902-1540	Mecklenburg-Vorpommern: Sülten	Mecklenburgische Seenplatte	20	Critically small population size. Within the same ecogeographic unit AgDZ matches genetic reserves quality criteria better than AgSL
AgSK	NRW-SK-20180818-1530	Nordrhein-Westfalen: Salzkotten	Westfälische Tieflandsbucht	30	The only representative of the ecogeographic unit, site management exists already
AgGOE	NI-GO-20150825-1000	Niedersachsen: Nörten-Hardenberg	Weser-Leinebergland	30	Within the same ecogeographic unit AgHI matches genetic reserves quality criteria better than AgGOE
AgHI	NI-HI-20150824-1600	Niedersachsen: Hildesheim	Weser-Leinebergland	30	Best representative of the ecogeographic unit
AgJX	NI-JX-20150724-1100	Niedersachsen: Jerxheim	Nördliches Harzvorland	30	The only representative of the ecogeographic unit
AgHES	HE-BS-20150925-1020	Hessen: Bad Salzhausen	Osthessisches Bergland	27	The only representative of the ecogeographic unit, site management exists already

Table 2. The marker name, the total number of analyzed plants, the number of distinct alleles, the polymorphic information content (PIC), observed (H$_o$) and expected heterozygosity (H$_e$) is presented.

Locus	Number of Individuals	Number of Alleles	Polymorphic Information Content (PIC)	Observed Heterozygosity (H$_o$)	Expected Heterozygosity (H$_e$)
ECMS01 [1]	790	7	0.733	0.291	0.769
ECMS11	790	9	0.774	0.523	0.801
ECMS13	790	8	0.193	0.066	0.203
ECMS02	790	8	0.434	0.128	0.502
ECMS23	790	8	0.384	0.168	0.442
ECMS39	790	13	0.803	0.167	0.826
ECMS06	790	18	0.774	0.205	0.799
ECMS09	790	8	0.457	0.101	0.556
Fn09 [2]	790	4	0.109	0.057	0.112
Fn100	790	6	0.042	0.005	0.042
Fn62	790	8	0.163	0.010	0.169
QC28 [3]	790	6	0.734	0.115	0.772
QC43	790	18	0.641	0.075	0.667
QC53	790	5	0.286	0.044	0.310
QC75	790	7	0.432	0.963	0.538
QC86	790	5	0.364	0.099	0.399

Detailed marker information was published for the [1] ECMS, [2] Fn series and for the [3] QC series by References [9–11].

The Hardy–Weinberg principle (HWP) provides the theoretical framework within which genetic variation has been analyzed by many research groups dealing with similar subjects. Therefore, the Chi^2-test was performed for each of the 432 combinations (27 occurrences × 16 markers) to test if the HWP holds for each of the combinations. 143 combinations were invariable. Out of the remaining combinations, 208 deviated significantly from HWP and only 81 combinations (about 19%) were in Hardy–Weinberg equilibrium (detailed results not presented here). Since a high number of combinations deviated from HWP, the genetic distance and genetic differentiation among the 27 occurrences was analyzed using the measure of Δ [43] and the software DifferInt [41].

Within the whole data set (790 individuals × 16 markers), 125 distinct alleles (excluding null alleles), 156 distinct single-locus types and 637 distinct multi-locus types were identified. Within occurrences, specific multi-locus genotypes were found to be duplicated between 2 and 18 times indicating self-fertilisation or preferential pairing within half- or full-sib families. The degree of inbreeding within an occurrence can be assessed by calculating the F$_{IS}$-index which is a measure of within population deficit of heterozygotes [44]. F$_{IS}$ ranged between −0.167 (occurrence AgG) and 0.558 (occurrence AgFRI) whereby a negative value indicates excess of heterozygotes and a positive value an excess of homozygotes within the occurrence. Five of the 27 occurrences show an excess of heterozygotes while 22 are characterized by an excess of homozygotes (Table 3).

Table 3. F$_{IS}$-values calculated over 16 markers for each of the 27 occurrences. The LabID is presented in bold letter type.

LabID	**AgBBG**	**AgBEN**	**AgBW**	**AgDA**	**AgDZ**	**AgEICH**	**AgFEH**
F$_{IS}$	0.043	−0.145	0.151	0.43	0.436	0.377	−0.016
LabID	**AgFRI**	**AgG**	**AgGOE**	**AgHEC**	**AgHES**	**AgHEU**	**AgHI**
F$_{IS}$	0.558	−0.167	0.363	0.231	0.441	−0.006	0.099
LabID	**AgHID**	**AgJX**	**AgKREM**	**AgOEH**	**AgROS**	**AgSEHL**	**AgSK**
F$_{IS}$	0.39	0.174	0.338	0.444	0.265	0.508	0.054
LabID	**AgSL**	**AgSUE**	**AgUW**	**AgWA**	**AgWU**	**AgZIE**	
F$_{IS}$	0.137	0.353	0.348	−0.08	0.49	0.14	

A Discriminant Analysis of Principal Components (DAPC) [45] was performed to determine the genetic structures within the sampled material (Figure 1). Genetic data from 757 of the 790 *A. graveolens* individuals that had been analyzed with 16 SSR markers were used for the calculation. These individuals correspond to 27 occurrences sampled in Germany (Figure 2). Thirty-three individuals having a null allele at any of the marker loci were excluded (see chapter Material and Methods). The DAPC revealed a much clearer separation of populations (Figure 1) than a single principle component analysis performed with DARwin [46] (data not shown). The first two axes plotted account for 46% of the observed variance. Individuals form interpretable clusters. They are arranged in two major groups and three offside occurrences AgG, AgUW and AgFEH corresponding to Baltic Sea coastal habitats. One major group is comprised of mainly coastal occurrences (AgWU, AgDA, AgHEU, AgHID, AgOEH and AgDZ) and appears more condensed (Figure 1, left part). This group also includes inland occurrences (AgSL, AgFRI, AgZIE, AGROS, AgBEN). The other major group of occurrences exhibits a greater range and harbours mainly inland occurrences (Figure 1, right part) (AgJX, AgHES, AgSK, AgBBG, AgSUE, AgHEC, AgBW, and AgWA). The coastal occurrence AgEICH, AgKREM and AgSEHL are interspersed in the second major group in the lower right part of the scatter plot. The genetic structure revealed by DAPC analysis is therefore only partly in accordance with the geographic origin of the samples. The correlation between genetic clustering and the geographic origin of individual plants was further investigated by the K-means ex nihilo clustering method. Based solely on the genetic data the clusters identified correspond to the individual occurrences except for only one cluster harbouring individuals from neighbouring inland occurrences AgHEC and AgSUE and another one with coastal occurrences AgFEH and AgHID (data not shown).

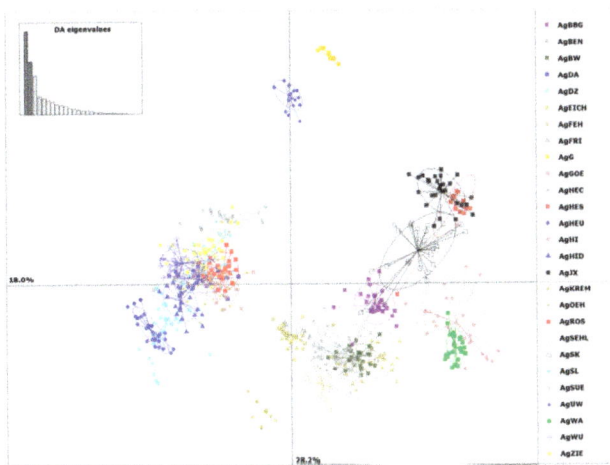

Figure 1. The scatter plot of the Discriminant Analysis of Principal Components (DAPC) shows the genetic structure within the collection of 27 occurrences of *A. graveolens* ssp. *graveolens*. Colors refer to the ecogeographic unit where the occurrence was sampled. ▮ = Weser-Aller-Flachland, ▮ = Nordmecklenburgisches Flachland, ▮ = Schleswig-Holsteinisches Hügelland, ▮ = Osthessisches Bergland, ▮ = Mecklenburgisch-Vorpommersches Küstengebiet, ▮ = Mecklenburgische Seenplatte, ▮ = Mittelbrandenburgische Platten und Niederungen, ▮ = Gäuplatten im Neckar- und Tauberland, ▮ = Weser-Leinebergland, ▮ = Mitteldeutsches Schwarzerdegebiet, ▮ = Elbtalniederung, ▮ = Nördliches Harzvorland, ▯ = Westfälische Tieflandsbucht, ▮ = Thüringer Becken und Randplatten.

Figure 2. Map showing the locations of the 27 occurrences. A black triangle marks the 15 Most Appropriate Wild Populations (MAWP) while a black dot marks the remaining 12 occurrences. The location names are given in Table 1.

The application of the genetic reserve conservation technique requires the choice of populations on the basis of the genetic representativity or uniqueness in relation to all analyzed populations [41]. In seed propagated species, under natural conditions, the parent population passes genes or gene complexes with their specific alleles to the next generation. Therefore, the recommendation of populations for conservation of a seed propagated species has to be based on the analyses of compositional genetic differentiation (Δ_j) at the gene pool level. Next to Δ_j, quality criteria for genetic reserves suggested by Iriondo et al. [32] were applied to identify MAWP (Table 1).

The measure Δ_j, marked in the snail diagram by the radius length of a sector, was calculated to assess the contribution of each occurrence to genetic differentiation. The result is summarized in Figure 3. The mean genetic differentiation Δ_{SD} is the average of the 27 radii and is marked by the circle in the diagram. The mean complementary genetic differentiation is $\Delta_{SD} = 0.3688$. AgHEC ($\Delta_j = 0.2513$) represents the genetic composition of the pooled remaining 26 occurrences best while the genetic composition of AgWA ($\Delta_j = 0.4776$) deviates from its complement most. The Δ_j values of the other 25 occurrences decreases in small steps clockwise from AgG to AgROS. The difference between the Δ_j-values of two occurrences depicted next to each other in clockwise sense is never larger than 0.034, i.e., less than 3.4% of the maximal possible sector radius length, and for most of the occurrences (20 out of 27) smaller than 1%. This means that neighbouring occurrences differ only slightly in their contribution to genetic differentiation.

The genetic reserve conservation technique includes by definition the active management of a MAWP. Knowledge of the reproductive system of the species is required to understand how site management interventions influence the genetic composition and the development of genetic diversity of a population at a specific site over time. Information on genetic bottlenecks caused by extinction and re-colonization events in the past can likewise improve the operation of genetic reserves. Site-specific selection pressure can change the distribution of alleles over populations within the species gene pool. The effect of selection and other forces influencing the genetic composition of populations show up as differences in the genetic composition among populations and are expressed as deviation from the HWP. The possible causes for deviations from the HWP can be investigated by permutation analyses [41].

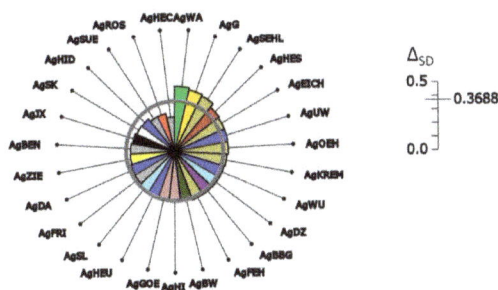

Figure 3. Snail diagram showing the differentiation between 27 occurrences of *A. graveolens* at the gene pool level. A sector represents one of the occurrences. The side length of a sector quantifies the contribution of each occurrence to differentiation. The grey circle marks the mean differentiation and the average of all 27 radii length. Colors refer to the ecogeographic unit where the occurrence was sampled. Identical colors were used in Figure 1.

In the first permutation analysis, all alleles at each locus were permuted among the individuals within populations. The permutation test was performed to investigate the influence of the reproductive system on the distribution of alleles within *A. graveolens* populations. In an ideal panmictic population alleles associate at single loci (homologous association) independent of the allelic type (fragment length). The observed mean differentiation increased from the gene pool ($\Delta_{SD} = 0.3688$), and mean single-locus ($\Delta_{SD} = 0.3771$, $p = 0.0$) to the multi-locus level ($\Delta_{SD} = 0.4001$, $p = 0.0$) of genetic integration (Table 4). The Δ_{SD}-value observed at the mean single-locus level of genetic integration is outside the range of the 95% confidence interval as is the Δ_{SD}-value observed at the multi-locus level, i.e., the observed values are not part of the majority of the Δ_{SD}-values calculated from data sets generated by 10,000 random permutations. At the mean single-locus, and multi-locus level the observed Δ_{SD}-value is smaller than 95% of all the Δ_{SD}-values generated by permutations. The mean differentiation at higher levels of genetic integration is less than could be expected from random gene association. The hypothesis that linkage equilibrium or HWE exists and that gene associations are created within populations independently of the allelic type at each locus is to be rejected [41].

The presence of a specific genotype within a population can either be incidental, the effect of selection within a population or the result of migration between populations. In the second permutation analysis all individual genotypes with their genetic types were randomly permuted among the occurrences. The hypothesis that forces that assign individuals to populations do this independently of their genetic types at a given level of genetic integration was tested. At all integration levels the observed Δ_{SD}-value was significantly higher than those resulting from 10,000 random permutations, i.e., the differentiation among occurrences was higher than could be expected from random association of genotypes to occurrences. In other words, individual genotypes are not randomly distributed across the 27 occurrences. If we define the 27 occurrences as subpopulations of meta-population, it can be concluded that a significant genetic differentiation between subpopulations exists. As the occurrences were sampled at large geographic distances (see Figure 2) this finding can rather be explained by a site-specific genotypic selection and adaptational differentiation than by migration.

Table 4. Permutation analysis of alleles over individuals within occurrences and of all individual genotypes among occurrences. The values are based on 10,000 permutations for all 16 loci. The observed Δ_{SD} value is shown for the three integration levels. The calculated minimum and maximum Δ_{SD} values, the 95% confidence interval as well as the permutation (p) values are shown. The symbols **-> (upper part of the distribution) and **<- (lower part of the distribution) indicate that fewer than 1% and more than 99% of the permutations yielded Δ_{SD} values that are significantly equal or greater than the observed distance. ** = $p \leq$ 0.01.

Permutations	Observed Δ_{SD} Values	Min	0.95 Confidence Interval	Max	p Value
Alleles over individuals within occurrences					
Gene pool level	0.3688		not affected		
Mean single-locus genotypes	0.3771	0.3782	0.3790; 0.3810	0.3821	1.0000 **<-
Multi-locus genotypes	0.4001	0.4060	0.4066; 0.4080	0.4085	1.0000 **<-
All individual genotypes among occurrences					
Gene pool level	0.3688	0.0960	0.1001; 0.1102	0.1163	0.0000 **->
Mean single-locus genotypes	0.3771	0.1092	0.1129; 0.1227	0.1286	0.0000 **->
Multi-locus genotypes	0.4001	0.2252	0.2284; 0.2359	0.2415	0.0000 **->

3. Discussion

Microsatellite markers have been widely applied to study patterns of genetic diversity of a species and to select and recommend populations for conservation purposes [47,48]. We succeeded to identify 16 polymorphic microsatellite markers. This marker set was used to characterize the genetic diversity of 27 *A. graveolens* ssp. *graveolens* occurrences. In the present study, the genetic distance measure Δ was used to quantify the contribution of each of the 27 occurrences to differentiation. The occurrence AgHEC proved to be most suitable to represent the genetic diversity of the remaining 26 occurrences and was identified as MAWP. The genetic composition of AgWA differed most from its complement. This specific diversity pattern can be interpreted as site-specific adaptation which makes AgWA a candidate for gene conservation measures. For several reasons, further 13 occurrences were selected as candidate MAWP. Microsatellite markers are very well-suited to assess the impact of selection and other evolutionary forces on the structure of genetic diversity of a species. However, they do not detect adaptive trait variation. In order to increase the chance of capturing adaptive trait variation, occurrences from each of the ecogeographic units were chosen. If several occurrences originated from the same ecogeographic unit, the one with the largest population size was nominated as MAWP. Preference was also given to sites already managed by local conservation agencies (see Table 1, LabID in bold letter type and Figure 2, black triangles). However, the nomination of 15 occurrences by no means devalues the remaining 12 occurrences. According to the definition of a genetic reserve, the population will actively be managed to maintain or, if necessary, improve the status of the target population. Active management requires financial resources which are always limited. The nomination of a limited number of MAWP is a pragmatic approach and helps spending financial resources efficiently. If, for whatever reason, a genetic reserve gets lost, it can be replaced by one of the occurrences with a similar Δ_j-value. This pragmatic approach is feasible as the differences between the Δ_j-values of occurrences ordered on the snail diagram next to each other is often less than 1% (Figure 3).

The mean genetic differentiation within the set of 27 occurrences is lower than $\Delta_{SD} = 0.5$ ($\Delta_{SD} = 0$ signifies no differentiation, $\Delta_{SD} = 1$ signifies complete differentiation) at all levels of genetic integration (gene pool level: $\Delta_{SD} = 0.3688$, mean single-locus level: $\Delta_{SD} = 0.377$, multi-locus level: $\Delta_{SD} = 0.4001$) although plants were sampled over a north-south distance of 680 km between the location Hiddensee and the location Ubstadt-Weiher near Karlsruhe. The reason for the lower genetic differentiation in *A. graveolens* ssp. *graveolens* as compared to *Helosciadium nodiflorum*, a wild celery species sampled in a much smaller distribution area [49], is unknown. Future investigations of the long-distance distribution mechanisms may help explaining this observation.

A. graveolens ssp. *graveolens* occurs wild as a halophilous marsh plant along sea coasts which likely is the primary distributions area. With the development of farm land at the expense of forest land over the past 1400 years [50] a secondary distribution area, the inland salt marshes, evolved. It can be assumed that the species migrated from the primary distribution area and colonized inland salt marshes in Germany. Inland saline and wet biotopes either developed naturally around salty springs (occurrence AgSUE) or result from salt mining activities (AgWA) where the species can be found at wet places close to soil heaps left over from potassium mining. These sites are like geographically isolated islands.

The DAPC analysis showed two major groups of occurrences: a costal and an inland group. The coastal group contains occurrences from inland sites and vice versa (Figure 1). For the time being this pattern is difficult to explain. The role of past colonization events, as well as factors shaping genetic diversity of *A. graveolens* ssp. *graveolens* today (migration, site specific selection pressure leading to adaptational differentiation, demographic fluctuations as a result of site-specific management actions) is not yet understood and deserves future research.

In two cases genetic signatures of migration between occurrences might exist. The K-means ex nihilo clustering method sorted almost all individuals of AgSUE and AgHEC into a single cluster. The plants were sampled at about 20 km distance in a landscape named Sülzetal ("Sülze" means salty water, "Tal" means valley) which makes exchange of seeds carried by animals between both sites more likely. The distance between occurrences AgFEH and AgHID growing on the Baltic Sea islands Fehmarn and Hiddensee, respectively, is 140 km (Figure 2). Both islands are important resting places in the autumn and spring for migrating birds. The allocation of 15 AgHID individuals to the AgFEH cluster could be interpreted as genetic signature of long-range seed transportation by birds. However, as long as no information on seed dispersal mechanisms in wild celery is available, this assumption remains a working hypothesis.

Some sites are not managed (e.g., AgWA) while others are grazed extensively (AgSUE) or mowed in late summer to maintain or improve the conservation status of the occurrence. If such management interventions happen only sporadically, the populations will undergo strong demographic changes from one generation to the next. We interpret the excess of homozygotes and heterozygotes, respectively (Table 3), as the genetic signature of an unstable demographic development of the occurrences resulting in the foundation of smaller groups of full or half-sib plants scattered over the site. *A. graveolens* ssp. *graveolens* is a self-compatible, insect pollinated species [51] and preferential pairing between closely related plants is therefore possible. Out of the 27 analyzed occurrences, 22 showed excess of homozygotes. As indicated by the permutation analysis, gene associations are created within occurrences dependent of the allelic type at each locus (Table 4, upper part). Self-pollination or mating between closely related individuals is therefore a likely explanation of this finding.

Iriondo et al. proposed quality standards to support practitioners involved in the design and management of genetic reserve. The population size is an important quality and selection criterion and should be large enough to sustain the long-term population viability [32]. Estimates of the demographic and genetic minimum viable population (MVP) for *A. graveolens* ssp. *graveolens* could not be found in the literature. However, animal genetic resources experts dealt with a similar problem already. Animal breeds are classified into four categories depending on their current effective population size whereby populations with an effective population size of $N_e > 1000$ individuals are considered not threatened. If N_e is within the range of 200 to 1000 individuals, populations are monitored and if the number of adult males sinks below 100 individuals, genetic conservation actions are initiated [52]. This approach can be transferred to plant genetic resources management programmes as it was derived from calculations of the increase of inbreeding depending on the allele frequency and the number of generations of random mating within populations of a given N_e and holds for any species. In order to meet the quality standards and as a first pragmatic recommendation for genetic reserve managers, the N_e of an *A. graveolens* ssp. *graveolens* population should not drop below a size of 1000 flowering individuals in a genetic reserve site. Sites like AgZIE with a population size of several hundred individuals fall already into this category while sites like AgWA do not. At AgWA interventions are required to enlarge the population area which will promote the spread of the occurrence and the increase of the population size. Although an effective population size of between 200 and 1000 individuals seems too high, it is achievable if the 15 recommended sites are managed accordingly. Future research projects could include occurrences from the whole European distribution area as to find occurrences in partner countries which genetically complement the suggested network of 15 sites for *A. graveolens* ssp. *graveolens* in Germany. By doing so, a European ex situ collection of wild *A. graveolens* ssp. *graveolens* accessions can be built representing a maximum of genetic diversity with a minimum of accessions.

Just today, when crop wild relatives are needed as sources of novel genetic variation more than ever before [53], the loss of species continues largely unrestrained [54,55]. Anthropogenic climate change has impact on plant species and may cause genetic erosion within species. Parmesan and Hanley [56] reviewed five global meta-analyses from long-term observational data on wild plant species and found that 44–65% of all investigated species show significant long-term change in phenologies, distribution, abundance and morphology. The changes correlated significantly with local or regional climate change patterns observed within the distribution area. To what extent climate change will jeopardize the long-term viability of specific species depends inter alia on the ability and speed of the species to migrate and colonize new sites [57]. In view of the relatively short period left, scientists call for the reinforcement of ex situ collecting activities to close gaps in genebanks and save germplasm for future generations [58]. This approach is justified as we have apparently no choice but to store the genetic resources of plant species in deep freezers and to reintroduce the material when mankind has overcome the climate crisis or any other devastating event. Ex situ conservation is one option. The better option is to develop and implement a conservation strategy which combines ex situ preservation of germplasm in genebanks with the maintenance of populations of the species in their natural habitat "in situ" where they have developed their specific traits [59,60]. There are two main differences between the in situ and the ex situ conservation approach. The management of a crop wild relative population in genetic reserves, such as the ones to be suggested for the establishment of *A. graveolens* ssp. *graveolens*, by definition includes the maintenance of favourable habitat conditions so that the target species can reproduce, adapt to changing growing conditions and evolve. The genetic reserve conservation technique [31] combines the best components of the ex situ and in situ conservation strategy and is a core element of the concept for in situ conservation of crop wild relatives in Europe endorsed by the steering committee of the European Cooperative Programme for Plant Genetic Resources (ECPGR) [30,61].

As compared to ex situ conservation, in situ conservation is a rather young research domain. If the distribution ranges of species shift as predicted it obviously will have important practical consequences for the genetic reserve management [57]. Climate change may also negatively affect *A. graveolens* ssp. *graveolens* within the European distribution area in the next 50 years [62,63]. A forecast of possible impacts of climate change on wild celery populations still is constrained by a shortage of high quality distribution data which can be used for modelling [64]. Furthermore, the reaction of plant species to climate change is more complex than thought before. Therefore, Parmesan and Hanley [56] called for the development of coordinated experiments across networks of field sites to better understand the response of species. The networks are not just a set of sites of plant populations but also a network of institutions and people who can record long-term data series on changes in phenologies, distribution, demography as well as genetic changes. Data required to forecast changes based on causal relations between the characteristics of a species and climate factors with a given statistical probability can be recorded by species-specific genetic reserves networks such as the one suggested for wild celery.

The flow chart (Figure 4) illustrates how the establishment of a genetic reserve network in Germany can be organised [23]. It summarizes experiences of a model and demonstration project. The whole process starts with the gathering of distribution data which are combined and processed (Steps 1 and 2). The existence of the plant species at selected sites is then confirmed by experts in the various Laender, material for genetic analysis is sampled and analyzed (Steps 3–5), the results and recommendations are discussed with stakeholders, the genetic reserve sites planned and the genetic reserve is finally established (Steps 6–8). Partners willing to contribute to the management of a specific genetic reserve site sign an agreement. Finally, their contribution to the management of a genetic reserve is acknowledged in a certificate issued by the Ministry of Food and Agriculture. The process can be divided into an information phase (Steps 1–5) and participatory project planning phase (Steps 6–8). Right from the beginning potential stakeholders are involved in the project planning and implementation, and are kept informed. A structured and open-result discussion with stakeholders is an important element of the procedure.

Next to these biological and scientific advantages of the in situ conservation approach the role of genetic reserves in public awareness building cannot be overemphasized. Genetic reserves are conservation projects that require the support and active collaboration of local people without which a genetic reserve can neither be established nor maintained over a longer period. Local communities can easily be convinced that genetic reserves contribute to food security and serve people. Genetic reserves are therefore not only a means of maintaining the environmental conditions under which economically important plant species can continue to evolve. The planning process is also a means for raising public awareness regarding the relationship of climate change, nature and species conservation, breeding progress and food security.

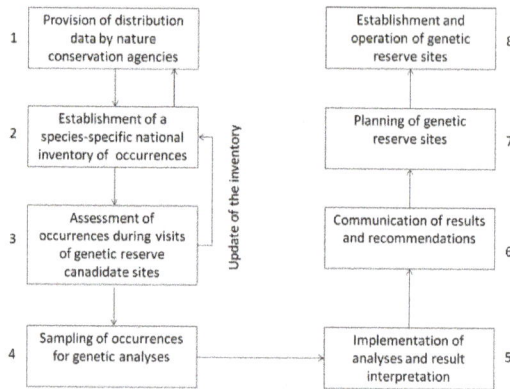

Figure 4. The flow diagram illustrates the procedure of establishing species-specific genetic networks in Germany.

4. Material and Methods

Distribution data of *A. graveolens* ssp. *graveolens* were provided by nature conservation agencies in Germany. A list of 690 sites where the species was sighted over the past decades was compiled. Sites located in differing ecogeographic units (as defined by Reference [42]) were tagged in the project database in order to increase the chance of capturing a wide range of adaptive trait variation. Among the tagged sites those with information on the population size or on social factors such as the potential support of the genetic reserve concept by local people were chosen. Altogether 78 sites located in nine of the 16 German Federal Laender were finally visited and assessed by contractors. The presence of *A. graveolens* was confirmed for 64 sites in the year 2015. Based on the contractor's report (plant number higher than 30 individuals, suitability of the site as genetic reserve) and representation of the highest possible number of differing ecogeographic units, 27 sites were chosen, and leaf samples taken in 2016. At 24 sites up to 30 individuals were sampled as planned while only 27, 23, and 21 individuals could be sampled at the site AgHE, AgDA, and AgSL, respectively. About half of the samples were taken in the coastal region of the Baltic Sea and in the centre of Germany, respectively. The geographic origin of the material and the sample codes are given in Table 1.

In this paper we denote a group of plants growing at a specific site as "occurrence" as it is not known whether any of the occurrences (generally called subpopulation) or the ensemble of all studied occurrences (generally called meta-population) fulfils the definition of the term population given by Kleinschmit et al. [65]. Instead of postulating the existence of populations we preferred to use an unbiased term which is also used by the Global Biological Information Facility (GBIF) in the sense of a sighting or sampling of a species [66].

Genomic DNA was extracted from 5 mg dried leaf material after vigorous homogenization in a mixer-mill disruptor with the innuPREP Plant DNA Kit (Analytik Jena AG, Jena, Germany) using lysis solution SLS according to the manufacturer instruction. DNA amplification was carried out in a total volume of 10 μL. The labelling of PCR products in one reaction was performed with three primers. The PCR mix contained 25 ng template DNA, 1.5 mM $MgCl_2$, 200 μM of each dNTP, 0.05 μM of a sequence-specific forward primer with M13 tail at its 5′ end, 0.17 μM of a sequence-specific reverse primer, 0.035 μM of the universal fluorescent-labelled (dye: D2, D3, D4) M13 primer and 0.5 U Taq DNA polymerase. Therefore, a multiplexing could be performed depending on the used marker. A touch-down PCR profile (TD 58–52 °C) was generally used as described by Nachtigall et al. [67]. The PCR products were separated and detected by using a capillary electrophoresis GeXP Genetic Analysis System (GenomeLab™, AB Sciex Germany GmbH, Darmstadt, Germany). Fragment sizes were determined and documented in a database developed by Enders [68]. The analysis of a probe was once repeated in case of the absence of a fragment. If the fragment remained absent, the allele was recorded as a null allele.

Thirty-eight published SSR markers [9–11] were screened to identify polymorphic markers. To this end a total of 30 plants (three populations of different geographic origin × 10 individuals) were used. Sixteen polymorphic markers were detected and used to genotype up to 30 individuals per occurrence giving a set of 12,640 data points. The markers were derived from coding DNA-sequences and should contain functional information [10]. Since none of 16 SSR markers have been mapped it is not known how well linkage groups are represented. The complete set of SSR marker data is available as open access document [69]. The SAS ProcAllele (SAS 9.4) procedure was used to calculate descriptive genetic parameters (Table 2). For each occurrence/SSR marker combination the deviation from the Hardy–Weinberg principle (HWP) was tested with the Chi^2-test ($p = 0.05$) using SAS ProcAllele and the result either indicated as HWE (in equilibrium) or HWD (in disequilibrium). The index F_{IS} was calculated using FSTAT 2.9.3 [44] to assess the excess of heterozygotes or homozygotes in a given occurrence.

In order to investigate the genetic structures within the research material, a Discriminant Analysis of Principal Components (DAPC) [45] was performed by using R version 3.5.1 [70] and the package adegenet version 2.1.1 [71]. The DAPC applies a multivariate method that maximises between-group and minimizes within-group components of genetic variation. A Structure [72] input file carrying genotypic information and population assignments was converted into a genind object using the read.structure function. A DAPC analysis was subsequently performed using the dapc function with 28 PCs explaining 89% of the total variance and 5 retained discriminant functions. The scatter.dapc function was employed to visualize individuals and clusters. In addition, K-means clustering was used by applying the function find.clusters to infer 27 genetic clusters from the genetic data ex nihilo without prior population information.

The genetic distance and genetic differentiation were calculated using the measure Δ [43]. The measure Δ is free of model assumptions such as the presence of large, random mating populations in Hardy–Weinberg equilibrium. The measure Δ ranges between 0 and 1. It can be used to calculate the complementary compositional differentiation within a set of several populations whereby Δ_j is the contribution of the jth populations to genetic differentiation. Δ_j is the genetic distance of the jth population to the pooled remainder ("the complement"). Δ_{SD} quantifies the average degree to which all populations differ from their complements. The computer programme DifferInt [41] was used for the statistical analysis of the data set. Only a few null alleles were detected, and 33 individuals were excluded from the data set prior to running the computer programme DifferInt. This computer programme is applicable to a set of co-dominant marker data without null alleles. Null alleles can be included in the analysis, but then results must be interpreted as phenotypic differences [73]. A significant advantage of this measure is that it allows the analysis of patterns of genetic differentiation in a set of occurrences at different levels of genetic integration, whereby the term "genetic integration" designates the arrangement of alleles into single-locus diplotypes and genes with their specific alleles into multi-locus genotypes [40]. The differentiation measure was obtained for three levels of genetic integration, namely the gene pool level, i.e., all alleles at one or more loci and the single-locus genotypes, the mean single-locus level. These two are characterized by locus and allelic state. The highest level of genetic integration is the multi-locus level. The multi-locus genotypes (individual plants) are characterized by the allelic states at all loci [41].

The differentiation patterns can be visualized by a differentiation snail, which is a pie-like chart. A sector of the chart represents one of the occurrences. The radius of its sector equals the contribution of this occurrence to differentiation. The sectors are arranged according to the radius lengths starting with the largest radius at 12:00 h. The sector with the second largest radius is placed to the right followed by the remaining in decreasing order depending of the individual radius length. The circle in the differentiation snail (Figure 3) is equal to the weighted mean of the sector radii and marks Δ_{SD}. The Δ_{SD}-value is also shown on the bar next to the snail graph. Occurrences representing the

Agriculture **2018**, *8*, 129

complement perfectly are characterized by $\Delta_j = 0$ while occurrences sharing none of the genetic types with the complement in common are indicated by $\Delta_j = 1$ [41]. Two different kinds of occurrences are of special interest for genetic conservation measures. The occurrence with the lowest contribution represents the composition of its complement best while the occurrence with the highest value has a genetic composition deviating from its complement [41,43].

The experimental data were used to derive information of the likely causes of the observed genetic differentiation. To this end ten thousand new data sets were generated by random permutation of all genes (alleles) at each locus among the individuals within each occurrence using DifferInt. To test the hypothesis that forces within occurrences create gene associations in individuals at a given level of integration do this independently of the allelic type at each locus, the genes within each population were randomly permuted among the individuals within each occurrence. If the p-value is of the observed differentiation is exceptionally small, i.e., smaller than $p = 0.05$, the hypothesis is rejected.

The term "gene association" denotes the non-random combination of alleles to genotypes at single loci (homologous association) or the non-random combination of single-locus genotypes to multi-locus genotypes (nonhomologous association) [41]. If alleles of a gene as well as different genes with their alleles associate non-randomly, there is linkage disequilibrium (LD) which has similarities to the HWD [74]. HWD exists when the frequency of homozygotes or heterozygotes within a population is higher or lower than what would be expected if the alleles and genes associate randomly. The first permutation test shows deviations from random distribution patterns at a specific significance level. Significant deviations indicate LD and HWD.

In a second permutation analysis all individuals together with their multi-locus types were randomly permuted among the populations. To test the hypothesis that forces that associate individuals with occurrences do this independently of their genetic type at a given level of integration the individuals were randomly permuted among the occurrences. If the p-value of the observed differentiation is exceptionally small, i.e., smaller than $p = 0.05$, the hypothesis is rejected. Deviations from random distribution patterns indicate the existence of population structures created by migration, selection, genetic linkage, the mating system or a combination thereof.

Author Contributions: M.B. coordinated and documented the field work. M.N. supervised the laboratory work, documented and processed the raw data. U.S. applied the various statistical software packages and assisted the data processing. L.F. supervised the work, interpreted the results and drafted the paper. All persons contributed to the writing of the paper.

Funding: This research was funded by the Bundesministerium für Ernährung und Landwirtschaft (BMEL) through the Bundesanstalt für Landwirtschaft und Ernährung (BLE), Bonn, Grant Number 2814BM110.

Acknowledgments: Information on the sites and status of the *A. graveolens* occurrences were recorded and reported by Joachim Daumann, Heiko Grell, Dietrich Hanspach, Klaus Hemm, Alexandra Kehl, Anselm Krumbiegel, Christof Martin, Ulrich Meyer-Spethmann, Heike Ringel who also provided leaf samples for genetic analysis. Ralph Engelmann developed a software tool for free which allowed us to produce colored snail diagrams. Peter Wehling and Lorenz Bülow reviewed an advanced draft of the paper. Lorenz Bülow performed the DAPC analysis and provided further valuable help. The research work would not have been possible without the excellent laboratory work of Petra Hertling. We are also grateful for the critical comments and suggestions of four anonymous reviewers and in particular to Peter Civan.

Conflicts of Interest: The authors have no conflicts of interest to declare.

References

1.   AGMRC (Agricultural Marketing Resource Center). Available online: http://www.agmrc.org/ (accessed on 16 February 2017).

2.   EU (European Union); EuroStat. Your Key to European Statistics. Available online: http://ec.europa.eu/eurostat/web/main/home (accessed on 16 February 2017).

3.   Schultze-Motel, J. *Rudolf Mansfelds Verzeichnis Landwirtschaftlich und Gärtnerisch Kultivierter Pflanzenarten*, 2nd ed.; Akademie Verlag: Berlin, Germany, 1986; pp. 999–1001, ISBN 3-05-500-164-8.

4. Solberg, S.Ø. Celery. In *Leaf Medicinal Herbs: Botany, Chemistry, Postharvest Technology and Uses*; Ambrose, D.C.P., Manickavasagan, A., Naik, R., Eds.; CAP International: Wallingford, UK, 2016; pp. 74–84, ISBN 9781780645599.

5. Barbour, J.D. Vegetable Crops. Search for Arthropod Resistance in Genetic Resources. In *Global Plant Genetic Resources for Insect-Resistant Crops*, 1st ed.; Clement, S.L., Quisenberry, S.S., Eds.; CRC Press: Boca Raton, FL, USA, 1998; pp. 171–189, ISBN 0-84932695-8.

6. Melchinger, A.E.; Lübberstedt, T. *Abschlussbericht zum GFP-Forschungsvorhaben ghg 1/98 (97 HS 044). Sortendifferenzierung und Verwandtschaftsuntersuchungen bei Feldsalat (Valerianella locusta L.), Radies (Raphanus sativus L.) und Knollensellerie (Apium graveolens L.) mit Hilfe PCR-Gestützter Genetischer Marker*; Final report of the GFP research project ghg 1/98 (97 HS 044). Variety differentiation and investigation of the relationships in corn salad (*Valerianella locusta* L.), radish (*Raphanus sativus* L.) and celeriac (*Apium graveolens* L.) using PCR-based genetic marker; Universität Hohenheim, Institut für Pflanzenzüchtung, Saatgutforschung und Populationsgenetik, Lehrstuhl Angewandte Genetik und Pflanzenzüchtung: Stuttgart, Germany, 2003.

7. Torricelli, R.; Tiranti, B.; Spataro, G.; Castellini, G.; Albertini, E.; Falcinelli, M.; Negri, V. Differentiation and structure of an Italian landrace of celery (*Apium graveolens* L.): Inferences for on farm conservation. *Genet. Resour. Crop Evol.* **2013**, *60*, 995–1006. [CrossRef]

8. Wang, S.; Yang, W.; Shen, H. Genetic diversity in *Apium graveolens* and related species revealed by SRAP and SSR markers. *Sci. Hortic.* **2011**, *129*, 1–8. [CrossRef]

9. Acquadro, A.; Magurno, F.; Portis, E.; Lanteri, S. dbEST-derived microsatellite markers in celery (*Apium graveolens* L. var. *dulce*). *Mol. Ecol. Notes* **2006**, *6*, 1080–1082. [CrossRef]

10. Fu, N.; Wang, Q.; Shen, H.-L. De Novo Assembly, Gene Annotation and Marker Development Using Illumina Paired-End Transcriptome Sequences in Celery (*Apium graveolens* L.). *PLoS ONE* **2013**, *8*, e57686. [CrossRef] [PubMed]

11. Fu, N.; Wang, P.-Y.; Liu, X.-D.; Shen, H.-L. Use of EST-SSR Markers for Evaluating Genetic Diversity and Fingerprinting Celery (*Apium graveolens* L.) Cultivars. *Molecules* **2014**, *19*, 1939–1955. [CrossRef] [PubMed]

12. EURISCO. The European Search Catalogue for Plant Genetic Resources. Available online: http://eurisco.ecpgr.org (accessed on 8 May 2018).

13. Spoor, W.; Simmonds, N.W. Base-broadening: Introgression and incorporation. In *Broadening the Genetic Base of Crop Production*; Cooper, H.D., Spillane, C., Hodgkin, T., Eds.; CAP International: Wallingford, UK, 2001; pp. 71–79, ISBN 0-85199-411-3.

14. Breiteneder, H.; Hoffmann-Sommergruber, K.; O'Riordain, G.; Susani, M.; Ahorn, H.; Ebner, C.; Kraft, D.; Scheiner, O. Molecular Characterization of Api g 1, the Major Allergen of Celery (*Apium graveolens*), and its Immunological and Structural Relationships to a Group of 17kDa Tree Pollen Allergens. *Eur. J. Biochem.* **1995**, *233*, 484–489. [CrossRef] [PubMed]

15. Gadermaier, G.; Egger, M.; Girbl, T.; Erler, A.; Harrer, A.; Vejvar, E.; Liso, M.; Richter, K.; Zuidmeer, L.; Mari, A.; et al. Molecular characterization of Api g 2, a novel allergenic member of the lipid-transfer protein 1 family from celery stalks. *Mol. Nutr. Food Res.* **2011**, *55*, 568–577. [CrossRef] [PubMed]

16. Vejvar, E.; Himly, M.; Briza, P.; Eichhorn, S.; Ebner, C.; Hemmer, W.; Ferreira, F.; Gadermaier, G. Allergenic relevance of nonspecific lipid transfer proteins 2: Identification and characterization of Api g 6 from celery tuber as representative of a novel IgE binding protein family. *Mol. Nutr. Food Res.* **2013**, *57*, 2061–2070. [CrossRef] [PubMed]

17. Ochoa, O.; Quiros, C.F. *Apium* wild species: Novel sources for resistance to late blight in celery. *Plant Breed.* **1989**, *102*, 317–321. [CrossRef]

18. Diawara, M.M.; Trumble, J.; Quiros, C.F.; Millar, J.G. Resistance to *Spodoptera exigua* in *Apium prostratum*. *Entomol. Exp. Appl.* **1992**, *64*, 125–133. [CrossRef]

19. Trumble, J.T.; Dercks, W.; Quiros, C.F.; Beier, R.C. Host plant resistance and linear furanocoumarin content of *Apium* accessions. *J. Econ. Entomol.* **1990**, *88*, 519–525. [CrossRef]

20. Trumble, J.T.; Diawara, M.M.; Quiros, C.F. Breeding resistance in *Apium graveolens* to *Liriomyza trifolii*: Antibiosis and linear furanocoumarin content. *Acta Hortic.* **2000**, *513*, 29–38. [CrossRef]

21. Constance, L.; Chuang, T.-L.; Bell, C.R. Chromosome numbers in *Umbelliferae*. V. *Am. J. Bot.* **1976**, *63*, 608–625. [CrossRef]

22. IUCN (International Union for Conservation of Nature and Natural Resources). The IUCN Red List of Threatened Species. Available online: http://www.iucnredlist.org/details/164203/0 (accessed on 8 May 2018).

23. Frese, L.; Bönisch, M.; Herden, T.; Zander, M.; Friesen, N. In-situ-Erhaltung von Wildselleriearten. Ein Beispiel für genetische Ressourcen von Wildlebenden Verwandten von Kulturarten (WVK-Arten). *Nat. Prot. Landsc. Plan.* **2018**, *50*, 155–163.

24. Berg, C.; Dengler, J.; Abdank, A.; Isermann, M. *Die Pflanzengesellschaften Mecklenburg-Vorpommerns und ihre Gefährdung*; Textband; Landesamt für Umwelt, Naturschutz und Geologie Mecklenburg-Vorpommern, Weißdorn-Verlag: Jena, Germany, 2004; ISBN 3-936055-03-3.

25. Preising, E.; Vahle, H.C.; Brandes, D.; Hofmeister, H.; Tüxen, J.; Weber, H.E. *Die Pflanzengesellschaften Niedersachsens—Bestandsentwicklung, Gefährdung und Schutzprobleme Salzpflanzengesellschaften der Meeresküste und des Binnenlandes*; Naturschutz und Landschaftspflege in Niedersachsen; Niedersächsische Landesverwaltungsamt—Fachbehörde für Naturschutz: Hannover, Germany, 1990; ISBN 392232150X.

26. Bettinger, A. Die Binnensalzstellen in Lothringen, im Saarland und in Rheinland-Pfalz. In *Binnensalzstellen Mitteleuropas. Internationale Tagung, Bad Frankenhausen, Germany, 8–10 September 2005*; Thüringer Ministerium für Landwirtschaft, Naturschutz und Umwelt: Erfurt, Germany, 2007; pp. 143–148.

27. Böttcher, H. Das EU-Life-Natur-Projekt, Erhaltung und Entwicklung der Binnensalzstellen Nordthüringens (2003–2008). In *Binnensalzstellen Mitteleuropas. Internationale Tagung, Bad Frankenhausen, Germany, 8–10. September 2005*; Thüringer Ministerium für Landwirtschaft, Naturschutz und Umwelt: Erfurt, Germany, 2007; pp. 54–62.

28. Herrmann, A. Binnensalzstellen in Brandenburg—Verbreitung und Zustand salzbeeinflusster Lebensräume. In *Binnensalzstellen Mitteleuropas. Internationale Tagung, Bad Frankenhausen, Germany, 8–10. September 2005*; Thüringer Ministerium für Landwirtschaft, Naturschutz und Umwelt: Erfurt, Germany, 2007; pp. 135–142.

29. Kell, S. Farmer's Pride. Conserving Plant Diversity for Future Generations. Available online: http://www.farmerspride.eu/ (accessed on 8 May 2018).

30. Maxted, N.; Avagyan, A.; Frese, L.; Iriondo, J.M.; Magos Brehm, J.; Singer, A.; Kell, S.P. *ECPGR Concept for in situ Conservation of Crop Wild Relatives in Europe*; European Cooperative Programme for Plant Genetic Resources: Rome, Italy, 2015.

31. Maxted, N.; Hawkes, J.G.; Ford-Lloyd, B.V.; Williams, J.T. A practical model for in situ genetic conservation. In *Plant Genetic Conservation: The in situ Approach*; Maxted, N., Ford-Lloyd, B.V., Hawkes, J.G., Eds.; Kluwer Academic Publishers: Dordrecht, The Netherlands, 1997; pp. 339–364, ISBN 0-412-63400-7.

32. Iriondo, J.M.; Maxted, N.; Kell, S.; Ford-Lloyd, B.V.; Lara-Romero, C.; Labokas, J.; Magos Brehm, J. Quality Standards for Genetic Reserve Conservation of Crop Wild Relatives. In *Agrobiodiversity Conserving: Securing the Diversity of Crop Wild Relatives and Landraces*; Maxted, N., Dulloo, M.E., Ford-Lloyd, B.V., Frese, L., Iriondo, J.M., Pinheiro de Carvalho, M.Â.A., Eds.; CAB International: Wallingford, UK, 2012; pp. 72–77.

33. Kell, S.; Maxted, N.; Frese, L.; Iriondo, J.M. In situ conservation of crop wild relatives: A strategy for identifying priority genetic reserves sites. In *Agrobiodiversity Conservation: Securing the Diversity of Crop Wild Relatives and Landraces*; Maxted, N., Dulloo, M.E., Ford-Lloyd, B.V., Frese, L., Iriondo, J.M., Pinheiro de Carvalho, M.Â.A., Eds.; CAB International: Wallingford, UK, 2012; pp. 7–19.

34. Manel, S.; Schwartz, M.K.; Luikart, G.; Taberlet, P. Landscape genetics: Combining landscape ecology and population genetics. *Trends Ecol. Evol.* **2003**, *18*, 189–197. [CrossRef]

35. Hodel, R.G.J.; Segovia-Salcedo, M.C.; Landis, J.B.; Crowl, A.A.; Sun, M. The report of my death was an exaggeration: A review for researchers using microsatellites in the 21st century. *Appl. Plant Sci.* **2016**, *4*. [CrossRef] [PubMed]

36. Arus, P.; Orton, T.J. Inheritance patterns and linkage relationships of eight genes of celery. *J. Hered.* **1984**, *75*, 11–14. [CrossRef]

37. Huestis, G.M.; McGrath, J.M.; Quiros, C.F. Development of genetic markers in celery based on restriction fragment length polymorphisms. *Theor. Appl. Genet.* **1993**, *85*, 889–896. [CrossRef] [PubMed]

38. Feng, K.; Hou, X.-L.; Li, M.-Y.; Jiang, Q.; Xu, Z.-S.; Liu, J.-X.; Xiong, A.-S. CeleryDB: A genomic database for celery. *Database* **2018**, 1–8. [CrossRef] [PubMed]

39. Li, M.Y.; Wang, F.; Jiang, Q.; Ma, J.; Xiong, A.-S. Identification of SSRs and differentially expressed genes in two cultivars of celery (*Apium graveolens* L.) by deep transcriptome sequencing. *Hortic. Res.* **2014**, *1*. [CrossRef] [PubMed]

40. Gillet, E.M.; Gregorius, H.-R. Measuring differentiation among populations at different levels of genetic integration. *BMC Genet.* **2008**, *9*, 60. [CrossRef] [PubMed]

41. Gillet, E.M. DifferInt: Compositional differentiation among populations at three levels of genetic integration. *Mol. Ecol. Resour.* **2013**, *13*, 953–964. [CrossRef] [PubMed]

42. Meynen, E.; Schmithüsen, J. *Handbuch der Naturräumlichen Gliederung Deutschlands/Unter Mitwirkung des Zentralausschusses für Deutsche Landeskunde, 1953–1962*; Bundesanstalt für Landeskunde und Raumforschung: Bad Godesberg, Germany, 1962.

43. Gregorius, H.-R.; Gillet, E.M.; Ziehe, M. Measuring Differences of Trait Distributions between Populations. *Biom. J.* **2003**, *45*, 959–973. [CrossRef]

44. Goudet, J. FSTAT (Version 1.2): A computer program to calculate F-statistics. *J. Hered.* **1995**, *86*, 485–486. [CrossRef]

45. Jombart, T.; Devillard, S.; Balloux, F. Discriminant analysis of principal components: A new method for the analysis of genetically structured populations. *BMC Genet.* **2010**, *11*, 94. [CrossRef] [PubMed]

46. Perrier, X.; Jacquemound-Collet, J.P. DARwin5, Version 5.0.158 Software. Available online: http://darwin.cirad.fr/darwin (accessed on 10 October 2010).

47. Fievet, V.; Touzet, P.; Arnaud, J.-F.; Cuguen, J. Spatial analysis of nuclear and cytoplasmic DNA diversity in wild sea beet (*Beta vulgaris* ssp. *maritima*) populations: Do marine currents shape the genetic structure? *Mol. Ecol.* **2007**, *16*, 1847–1864. [PubMed]

48. Petit, R.J.; El Mousadik, A.; Pons, O. Identifying populations for conservation on the basis of genetic markers. *Conserv. Biol.* **1998**, *12*, 844–855. [CrossRef]

49. Zander, M.(Albrecht Daniel Thaer-Institut für Agrar- und Gartenbauwissenschaften, Urbane Ökophysiologie der Pflanzen, Humboldt-Universität zu Berlin, Berlin, Germany). Personal communication, 2018.

50. Piorr, H.-P. Entwicklung der pflanzenbaulichen Flächennutzung in den Agrarlandschaften Deutschlands. In *Biologische Vielfalt mit der Land- und Forstwirtschaft?* Welling, P., Ed.; Reihe A: Angewandte Wissenschaft; Schriftenreihe des Bundesministeriums für Verbraucherschutz, Ernährung und Landwirtschaft: Bonn, Germany, 2002; pp. 127–135.

51. Klotz, S.; Kühn, I.; Durka, W. BIOLFLOR—Eine Datenbank zu biologisch-ökologischen Merkmalen der Gefäßpflanzen in Deutschland. In *Schriftenreihe für Vegetationskunde 38*; Bundesamt für Naturschutz: Bad Godesberg, Germany, 2002.

52. BMVEL (Bundesministerium für Verbraucherschutz, Ernährung und Landwirtschaft). *Tiergenetische Ressourcen*; Nationales Fachprogramm: Bonn, Germany, 2004.

53. Wehling, P.; Scholz, M.; Ruge-Wehling, B.; Hackauf, B.; Frese, L. Anpassung landwirtschaftlicher Kulturarten an den Klimawandel—Optionen aus Sicht der Züchtungsforschung. *J. Cultiv. Plants* **2017**, *69*, 47–50. [CrossRef]

54. Bilz, M.; Kell, S.; Maxted, N.; Lansdown, R.V. *European Red List of Vascular Plants*; Office of the European Union: Luxembourg, 2011.

55. EU (European Union). Halbzeitbewertung der EU-Biodiversitätsstrategie. *Natura* **2000**, *2016*, 39.

56. Parmesan, C.; Hanley, M.E. Plants and climate change: Complexities and surprises. *Ann. Bot.* **2015**, *116*, 849–864. [CrossRef] [PubMed]

57. Araujo, M.B.; Cabeza, M.; Thuiller, W.; Hannah, L.; Williams, P.H. Would climate change drive species out of reserves? An assessment of existing reserve-selection methods. *Glob. Chang. Biol.* **2004**, *10*, 1618–1626. [CrossRef]

58. Castaneda-Alvarez, N.P.; Khoury, C.K.; Achicanoy, H.A.; Bernau, V.; Dempewolf, H.; Eastwood, R.J.; Guarino, L.; Harker, R.H.; Jarvis, A.; Maxted, N.; et al. Global conservation priorities for crop wild relatives. *Nat. Plants* **2016**, *2*. [CrossRef] [PubMed]

59. CBD (Convention on Biological Diversity). *Convention on Biological Diversity*; Convention on Biological Diversity: Montreal, QC, Canada, 1992.

60. FAO (Food and Agriculture Organization). International Treaty on Plant Genetic Resources for Food and Agriculture. Available online: http://www.planttreaty.org/ (accessed on 16 February 2017).

61. ECPGR (European Cooperative Programme for Plant Genetic Resources). The ECPGR Concept for in situ Conservation of Crop Wild Relatives in Europe. Available online: http://www.ecpgr.cgiar.org/working-groups/wild-species-conservation/ (accessed on 20 June 2017).

62. CWRnl. Crop Wild Relatives (CWR) in the Netherlands. Available online: https://www.cwrnl.nl/nl/CWRnl/CWRspergewas/Apium-graveolens-L.htm (accessed on 8 May 2018).

63. Aguirre-Gutiérrez, J.; van Treuren, R.; Hoekstra, R.; van Hintum, T.J.L. Crop wild relatives range shifts and conservation in Europe under climate change. *Divers. Distrib.* **2017**, *23*. [CrossRef]

64. Hoekstra, R. (Centre for Genetic Resources, Droevendaalsesteeg 1, Building 107, 6708 PB Wageningen, The Netherlands). Personal communication, 2017.

65. Kleinschmit, J.R.G.; Kownatzki, D.; Gregorius, H.-R. Adaptational characteristics of autochthonous populations—Consequences for provenance delineation. *For. Ecol. Manag.* **2004**, *197*, 213–224. [CrossRef]

66. GBIF (Global Biological Information Facility). Available online: https://www.gbif.org/species/search (accessed on 8 August 2018).

67. Nachtigall, N.; Bülow, L.; Schubert, J.; Frese, L. Development of SSR Markers for the Genus *Patellifolia* (Chenopodiaceae). *Appl. Plant Sci.* **2016**, *4*, 8. [CrossRef] [PubMed]

68. Enders, M. *Entwicklung und Anwendung Molekularer und Informatorischer Werkzeuge zum Genetischen Monitoring bei Wildrüben*; Diplomarbeit im Fach Bioinformatik, Martin-Luther-Universität Halle-Wittenberg, Naturwissenschaftliche Fakultät III, Institut für Informatik: Halle, Germany, 2010.

69. Nachtigall, M.; Frese, L.; Bönisch, M. *Microsatellite Marker Data of Apium graveolens L. ssp. graveolens*; Open Agrar Repositorium: Quedlinburg, Germany, 2018.

70. R Core Team. *R: A Language and Environment for Statistical Computing*, version 3.5.1; R Foundation for Statistical Computing: Vienna, Austria, 2013; Available online: https://www.R-project.org (accessed on 8 August 2018).

71. Jombart, T. Adegenet: A R package for the multivariate analysis of genetic markers. *Bioinformatics* **2008**, *24*, 1403–1405. [CrossRef] [PubMed]

72. Pritchard, J.K.; Stephens, M.; Donnelly, P. Inference of population structure using multilocus genotype data. *Genetics* **2000**, *155*, 945–959. [PubMed]

73. Gillet, E.M. DifferInt Compositional Differentiation among Populations at Three Levels of Genetic Integration. User Manual. Available online: http://www.uni-goettingen.de/forstgenetik/differint (accessed on 17 November 2016).

74. Slatkin, M. Linkage disequilibrium—Understanding the evolutionary past and mapping the medical future. *Nat. Rev. Genet.* **2008**, *9*, 477–485. [CrossRef] [PubMed]

agriculture

MDPI

Article

Diversity of Cropping Patterns and Factors Affecting Homegarden Cultivation in Kiboguwa on the Eastern Slopes of the Uluguru Mountains in Tanzania

Yuko Yamane *, Jagath Kularatne and Kasumi Ito

International Corporation Center for Agricultural Education, Nagoya University, Nagoya-shi, Chikusa-ku, Furocho 464-8601, Japan; jagath@agr.nagoya-u.ac.jp (J.K.); kasumito@agr.nagoya-u.ac.jp (K.I.)
* Correspondence: yamane@agr.nagoya-u.ac.jp; Tel.: +81-52-789-4225; Fax: +81-52-789-4223

Received: 4 June 2018; Accepted: 29 August 2018; Published: 13 September 2018

Abstract: This study investigated what kind of diversities of cropping patterns observed in home gardens distributed on the eastern slopes of the Uluguru Mountains in central Tanzania, and how the diversity come into occurred. The major focus included the differences in ecological environment due to elevation, the impacts of the Ujamaa policy, and the characteristics of household members. Participatory observation with a one year stay in the study village was conducted to collect comprehensive information and to detect specific factors about formation of diversity cropping patterns of homegardens. The features of cropping patterns of the homegardens were assessed in an area distributed at altitudes of 650–1200 m. Many of the tree crops in this village originated from outside regions around the period of Tanzanian independence, and their cultivation spread throughout the village after the implementation of the Ujamaa policy. At present, village districts with many distributed homegardens with numerous tree crops are those that were confiscated from clans by the village government at the time of the Ujamaa policy and then redistributed to individuals. Cultivation of trees crops was very few at altitude of 900 m or more, because of cultivation characteristics of tree crops in this village were suitable for low altitude. In addition, since homegardens are considered to be abandoned for one generation only, their cropping patterns tended to easily reflect the ages and preferences of the members of the households living on them. The cropping patterns of the homegardens differed remarkably even between neighboring households owing to the cumulative effects of these multiple factors. Analysis using an inductive method—considering the background against which the phenomenon becomes evident after collecting the information from the target area in this manner—is thought to lead to an essential understanding.

Keywords: cropping patterns; ecology; homegardens; mountain agriculture; Tanzania

1. Introduction

Homegardens have a complex and diverse cropping pattern similar to the agroforestry where trees and herbaceous crops grow together. Homegarden agroforestry systems in the tropics are known for their structural complexity and diversity in crop and other plant species [1]. Studies focused on variations in the diversity of homegardens have revealed that the diversity is often not static, but changes in response to socio-economic dynamics [1–3]. Homegardens are a time-tested local strategy that are widely adopted and practiced in various circumstances by local communities with limited resources [4]. Consequently, homegardens should not be interpreted as a generic agro-forestry system with uniform diversity characteristics, but rather as involving different types of features with respect to species diversity [1,3]. Therefore, a prerequisite for obtaining a precise understanding of the relation between species diversity and homegarden sustainability is that a better insight is obtained into the different dimensions of homegarden diversity at spatial and temporal scales and at the level of both species and functional groups "[5]").

The current stance adopted by developing countries on agricultural development research employs a deductive approach to provide the necessary supports based on modern agricultural sciences, which in the phenomena such as cropping pattern are generally analyzed from the environmental or economic viewpoint [6,7]. Agricultural technical supports that conducted in developing countries attempt to disseminate agrarian technologies or new varieties, which were developed based on the modern agricultural science [7]. Population in developing countries has predicted to increase, especially, Africa's growth rate is said to be the highest. Therefore, from the viewpoint of food security, the devising methods, and methods appropriate for African agriculture are indispensable. However, it appears that the modern agricultural technologies are not widely accepted by the rural people in Africa, as they are not adapted to unique agriculture [8]. It is essential to consider the methods of rural development or agricultural technology support to be accepted [8]. In this regard, it is vital to provide necessary support based on the actual situation of the target area. However, at present, there is a little information to understand the essence of agriculture in rural areas in developing countries, especially in Africa.

Cultivation of the homegardens in Africa were focused from multiple point of view such as food condition [9], economical [10], environmental [11], and as a place to preserve diversity of crops [12].

This study aimed to provide information about a time-tested local strategy of homegardens that is obtained from different spatial and temporal dimensions and diversity to provide information to deepen the understanding of Africa's homegarden. Most previous studies on homegardens in eastern Africa have commonalities in their crops combination; most of them cultivate Musaceae crops such as banana (*Musa* spp.) or Ethiopian banana (*Ensete ventricosum* (Welw.) Cheesman) and coffee (*Coffea arabica/robusta* L.) [5,13]. Ethiopian banana and banana are cultivated as their staple food crops and coffee is cultivated as a commercial crop in each area. Even when they mention the diversity in cropping patterns of their homegardens, these combinations of crops are observed commonly within an area [14–18]. However, extreme differences were observed in cropping patterns in homegardens in Kiboguwa Vill.age located on the eastern slopes of the Uluguru Mountains in central Tanzania. In some homegardens, many kinds of tree crops—for example, African breadfruit (*Treculia Africana* Decne.), coco palm (*Cocos nucifera* L.), jackfruit (*Artocarpus heterophyllus* Lam.), cinnamon (*Cinnamomum zeylanicum* J.Presl), etc.—are cultivated together. On the other hand, in another homegarden, herbaceous crops such as maize or common bean or banana dominate. Sometimes, the types of crops which are observed in homegardens are quite different even among adjacent homegardens.

This paper focuses on the diversity of the cropping pattern observed in the homegardens in the target village. After clarifying how the cropping pattern is diverse, the factors behind the diversity in cropping patterns in homegardens in the target village are comprehensively considered and described; such factors include ecological difference due to the elevation of the mountain, the historical land tenure policy and differences in household members, etc. The method of participatory observation accompanied with long term stay is a powerful method as a method for collecting information unique to the targeted area, and in this research, various information of the target area is collected based on field work and various information tried to explain why planting is seen from multiple backgrounds peculiar to the area.

2. Materials and Methods

2.1. Study Area

Kiboguwa Village (E: 37°40′50–37°42′20, S: 6°59–7°00′35), located in central Tanzania in the Uluguru Mountains (Figure 1a,b), extends over approximately 55 km north to south and 30 km east to west. The eastern slopes of the mountains where the study was conducted receive abundant rainfall; hence, many tree crops and commercial crops such as banana are commonly cultivated in the villages located on these slopes [19]. The elevation of Kiboguwa Village is approximately 800 m (Figure 1b), suggesting that it is also central regarding the elevations of villages found around the eastern slopes (between 300 m and 1500 m). The villagers' homes and fields are located at elevations from 600 m to 1200 m; practically all of the land, including the slopes, is used for cultivation.

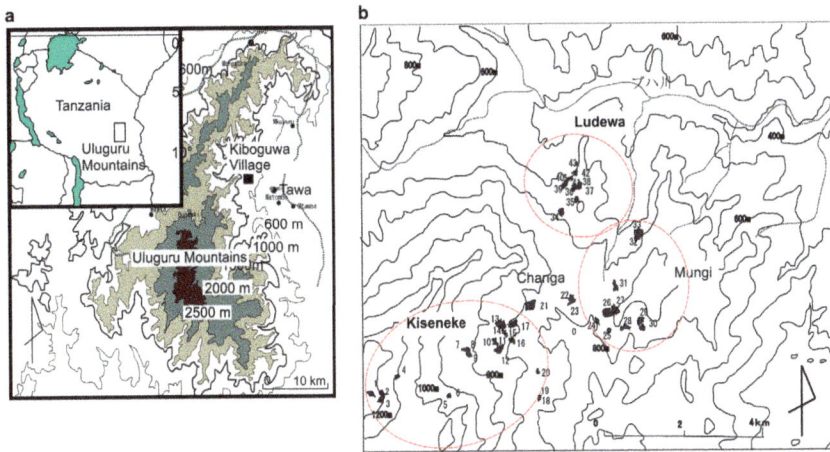

Figure 1. Location of Kiboguwa village in the Uluguru Mountains (**a**); and distribution of the 43 surveyed households in Kiboguwa village (**b**).

Two types of land use patterns were observed in Kiboguwa village, namely cultivated lands (primarily) and homegardens. The cultivated lands (*Kumugunda*, in the Luguru language) are used for the cultivation of staple food crops, such as maize (*Zea mays* L.), rice (*Oryza sativa* L.), and cassava (*Manihot esculenta* Crantz) [20]. Homegardens are called *Ditzulala* by the local people in Kiboguwa village. According to the villagers, this corresponds to the word *Jalala* in Swahili. However, *Jalala* refers only to a place where refuse from a house is discarded, whereas *Ditzulala* is used to refer to a homegarden. Currently, in a *Ditzulala*, spice crops such as cinnamon, pepper (*Piper nigrum* L.), and cardamom (*Elettaria cardamomum* (L.) Maton) grow together with tree crops such as African breadfruit, coconut, and coffee densely distributed around houses, creating a thick forest landscape. The villagers also plant herbaceous crops, including banana, maize, and common bean (*Phaseolus vulgaris* L.). Another land use pattern is the cultivated lands (*Kumugunda*, in the Luguru language) which are used for the cultivation of staple food crops—such as maize, rice, and cassava—which are distributed outside the homegardens.

In the Uluguru Mountains, the present social structure is centered on the Matrilineal clans of the Luguru people who live throughout the mountain range and greatly influence the land tenure in the region. According to the customary land tenure system of these people, male and female clan members have equal usufructuary rights from birth to death to cultivate the clan's land [21]. However, individuals are not recognized as owners of the land, and they do not have the property rights. Conversely, the ownership rights for perennial crops that last for many years, such as tree crops, belong to the person who planted them. This scenario creates the possibility of inheritance of the perennial crops as the property of children who are not members of the land's clan. Therefore, people with usufructuary rights for the land owned by their clan may differ from the owners of tree crops planted on that land. Thus, under the original land tenure system, tree crops and other perennial crops are recognized as difficult crops to plant on the clan's land due to the need to first obtain the consent of the clan members [21]. Nonetheless, at present, many tree crops that form a thick forest landscape are noted in the homegardens from the study region, but growing tree crops—i.e., perennial crops—are difficult to find on clan land [22]. In the Kiboguwa village, tree crops or perennial crops (including banana) are rare on the *Kumugunda*. On the other hand, the cultivation of perennial crop and tree crops has been observed in most homegardens of the village [20].

Recent studies of the Kiboguwa village already mention the role of the homegardens in the food security of the staple food crops [20] and the diversity of cropping patterns that was focused on the commercial crops [23]. There were many crops of secular life in commodity crops. Spice crops such as cinnamon and cardamom had a limited harvesting period, so they were used as a product crop to earn extra money. On the other hand, since bananas are obtained throughout the year, they were used as a stable income source. Bananas at the residence site were sometimes consumed boiled, but the earnings obtained from banana sales also played an important role in complementing self-sufficient crops [20]. This was caused by the low and insufficient productivity of the two main types of cereals, maize and rice, which were produced on the slopes outside the homegardens. Many households have been sold bananas grown in homegardens to compensate for cereal food shortages. Therefore, as a commercial crop, it was found that bananas play a vital role in achieving food security for the villagers [20].

According to the 2002 census, the population of Kiboguwa Village was 1402 (225,857 people in the Morogoro Region according to the 1988 census; in 2002, it reached 263,012 people; and in 2012, 286,248 people). The annual rate of population increasing from 1988 to 2002 was 1.17%, whereas from 2002 to 2012 it reached 0.88%, indicating that the population hardly increased. Therefore, a dramatic increase in the number of village households due to the population increase might not have occurred during the survey period (2004 to 2007) until 2013 (Data source: http://www.nbs.go.tz/). The survey includes seven village districts that cover the slopes of the ridge running through the center of the village at elevations of 800 m to 1400 m as well as the valleys. The difference in elevation is the largest in the Kiseneke village district. Between the Changa village district and the Buha River, Nbure, Ludewa, and Mungi village districts are located on the north side. The Mungi village district occupies an extensive range along the boundary with the tower of a neighboring village situated to the south.

2.2. Participatory Survey

A survey of participant observations accompanied by a long-term stay in the study area was carried out. The participant observation method was employed for the data analysis in this empirical research. This method is commonly used in anthropological studies [24]. The author lived with the farmers at their houses to obtain first-hand information and to understand actual situations on such factors as their lifestyle, food consumption, agricultural practices, etc. The author (Yamane) resided in Kuboguwa village for a total of one year as follows: three months from July to October 2004; four months from June to September 2005; one month in December 2005; two months from December 2006 to January 2007; and two months from June to July 2008. The cropping patterns of the study area and the factors—such as the history of homegardens, geography, climate, land use policies, and cultural behavior on the land use of this area—which were thought to affect the diversity of cropping patterns in the homegardens were studied. In this study, we conducted questionnaire surveys to collect quantitative data. The questionnaire was structured which corresponds to the information on the actual condition of the target area based on the qualitative information obtained from the participant observation. The causal relationship between the phenomena such as cropping pattern of the homegardens and the factors influencing on it were discussed. The author revisited the study village in December 2017, and it was confirmed that the agricultural landscape had not changed drastically.

2.3. Measurement of Cropping Patterns

A total of 254 households were identified among the four village divisions, via images captured by the QuickBird satellite sensor on 4 October 2005. A survey questionnaire was carried out in September 2005 among 43 randomly selected households in the four divisions of Kiboguwa village, to gather information about the homegarden. The number of households surveyed in the village divisions was

23 in Kiseneke, 3 in Changa, 10 in Mungi, and 10 in Ludewa. From December 2006 to January 2007, the same survey questionnaire was applied for a total of 84 households. The village council members in four districts were questioned regarding how they obtained the land for home gardening.

The location, size, and elevations of the homegardens were measured using a geographical information system (GPS; eTrex Legend; Germin Ltd. (Olathe, KS, USA)) in July 2005. A total of 43 households distributed within four village districts—20 households from Kiseneke village district, 10 households from each of Mungi and Ludewa village districts, and 3 households from Changa village district—were surveyed and the varieties of crops and the number of each crop were recorded separately. At the same time, a schematic diagram of the homegardens for these 43 households was developed (Figure 2). The cropping patterns were reproduced by using computer software, ArcView3.1 (ESRI, Redlands, CA, USA) to measure the exact size of the area of each crop variety. The heights of conventional tree crops were measured using a gauge.

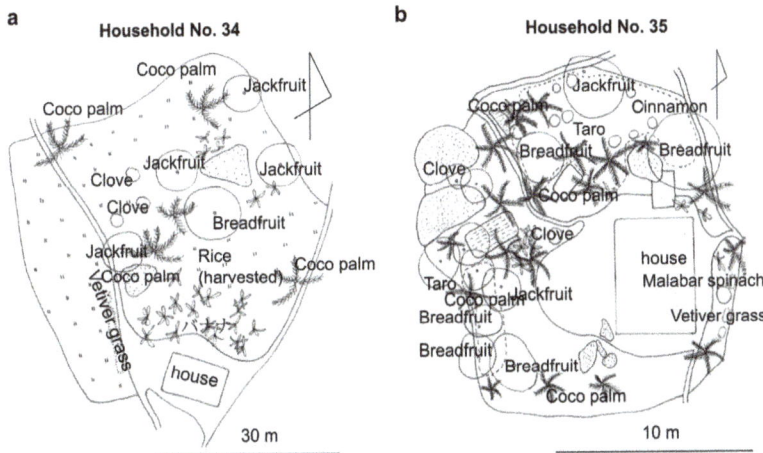

Figure 2. Schematic diagrams of the cropping patterns among neighboring homegardens (Ludewa).

The quantity of each crop variety was determined by using the schematic diagram that reflected the number of individual plants for herbaceous crops and the area size covered by each tree. The number of different plants such as maize and cassava could not be counted individually due to the high density of plants. In this regard, the quantity of the plants grown was expressed morphologically. The vine-like plants, such as sweet potato, were estimated by the ratios of the areas occupied by those plants. The sizes of the planted area of the crops mentioned above were obtained using the GPS data. The complexity of cropping patterns was expressed using Simpson's diversity index [25] and the value of the exponent ($D = (\Sigma sin/N)^2$) of Simpson were calculated.

In one home garden, there are many cases where a nuclear family, that is, a family mainly consisting of a couple and unmarried children live. The heads of the household of the four village districts were interviewed to obtain personal information such as age, place of birth, marital status, and the way in which they came to own the homegardens. In the case of the heads not knowing such information, other family members were interviewed. Therefore, in this paper, the husband of the nuclear family was regarded as 'household head'. In the cases where the male head of household was not present, the head woman of the household was the eldest female. Thus, the data were collected covering all households distributed in the four village districts.

2.4. Measurements of Temperature at Different Elevations

In September 2005, humidity and the temperature at elevations of 1050, 750, and 650 m in Kiboguwa Village were measured using metrological equipment (HOBO Pro RH, Klima Tech Co., Ltd., Tokyo, Japan).

2.5. Collection of Historical Information of Homegarden Crops

Some of the village elders were interviewed from July to October in 2004 to gather information such as the transition in agriculture, crop varieties, and the history of the introduction of tree crops in particular. Interviewed people consisted of 12 clan heads, and they were between 60–80 years old. Th elders were interviewed to confirm, to support, and to understand further detailed information of the previous surveys.

2.6. Questionnaire Survey

From December 2006 to January 2007, the same questionnaire survey was conducted for total 84 households; 21, 23, 18, and 22 households in the village divisions of Kiseneke, Changa, Mungi, and Ludewa, respectively. In the questionnaire of 2006 and 2007, some questions were included to obtain information about family member of the households and methods to get cultivated land and homegardens.

A total of 254 households were detected in the four village divisions when it was counted via the satellite image of Quick bird, which was taken on 4 October 2005. Our sample size was 84 households. Therefore, in this survey, approximately 33% households in four village divisions were targeted.

3. Results

3.1. Diversity of the Cropping Patterns in the Village Homegardens

The homegardens of Kiboguwa village presented extremely diverse cropping patterns. We classified the cropping patterns using five elements: (1) homegarden area; (2) ratio of crop cultivated area; (3) the varieties of crops planted in the cultivated area; (4) heights of the tree and herbaceous crop varieties; and (5) the diversity index which shows the crop diversity.

The homegarden area and the ratio of cultivated area: The area of homegardens belonging to the 43 households surveyed varied largely. The widest and narrowest homegarden areas ranged between 1.84 and 0.15 acres respectively, with an average of 0.66 acres. However, many households had sufficiently wide homegardens for cultivation (Table 1).

The range of the cultivated area ratios for four households (numbers 5, 18, 19, and 43) with a cultivated area of around 0.2 acres ranged from 39.9% to 82.9%. A household (number 38) with a homegarden smaller than the average (0.32 acre) planted crops on only 20% of its total area; in another household (number 27) with a large homegarden of 1.4 acres, the crops occupied nearly 80% of its total area. Thus, the ratios of cultivated area were thought to be not necessarily influenced by the size of the homegarden area. Therefore, the cropping pattern consisted of a characteristic of the relative use of the space of homegardens, and varied according to four factors: (i) ratio of crop cultivated area in homegardens; (ii) the varieties of crops planted in the crop cultivated area; (iii) heights of the tree and herbaceous crops varieties; and (iv) the diversity index. Based on those characteristics, we classified the cropping patterns in the homegardens of the 43 households.

Table 1. Information on the four types of cropping patterns.

Household No.	Cropping Pattern	Village District	Altitude (m)	Area (Acre)	Percentage of cropping Area (%)	Index of DIVERSITy *	Number of Crop Species			(3a) The Percentage of Area Occupied by Tree Crops			Tall Herbaceous Crop (%)	Medium Herbaceous Crops (%)	(3b) The Percentage of Area Occupied by Herbaceous Crops	
							Total (Spieces)	Tree Crops (Spieces)	Herbaceous Crops (Spieces)	Tall Tree (%)	Short Tree (%)	Total (%)			Short Herbaceous Crops (%)	Total (%)
35	Type I	Ludewa	658	0.37	53.2	0.59	13	6	6	89.4	3.7	93.1	5.4	0.5	1.0	6.9
39	Type I	Ludewa	658	1.00	29.5	0.77	13	9	4	71.2	16.8	88.0	4.5	4.0	3.5	12.0
27	Type I	Mungi	734	1.45	79.5	0.94	13	10	3	64.0	22.6	86.6	9.3	4.1	0.0	13.4
26	Type I	Mungi	736	1.38	54.3	0.78	13	10	3	80.4	3.9	84.3	11.6	1.6	2.5	15.7
22	Type I	Changa	810	0.58	73.7	0.80	16	9	7	66.7	14.8	81.5	13.2	5.3	0.0	18.5
43	Type I	Ludewa	640	0.24	39.9	0.73	8	6	2	73.3	8.1	81.4	18.6	0.0	0.0	18.6
28	Type I	Mungi	700	0.85	85.1	0.76	13	7	6	56.1	23.5	79.6	8.9	2.8	8.7	20.4
38	Type I	Ludewa	630	0.32	44.6	0.73	11	7	4	68.3	5.8	74.0	5.7	20.2	0.0	26.0
37	Type I	Ludewa	664	0.99	51.8	0.67	12	10	2	59.5	9.9	69.4	30.6	0.0	0.0	30.6
31	Type I	Mungi	660	0.80	70.3	0.91	19	11	8	66.1	3.0	69.1	23.0	5.3	2.5	30.9
29	Type I	Mungi	771	0.96	78.5	0.68	9	4	5	15.0	49.9	64.8	1.6	20.3	13.2	35.2
23	Type I	Changa	791	0.80	58.0	0.69	11	5	6	60.1	1.1	61.2	25.0	9.5	4.4	38.8
		Averadge	704	0.81	59.9	0.75	12.6	7.8	4.7	68.6	10.3	77.8	14.2	4.8	2.1	22.2
42	Type II	Ludewa	635	0.49	78.4	0.64	8	5	3	19.4	36.4	55.8	7.8	0.0	36.4	44.2
41	Type II	Ludewa	649	0.39	26.0	0.58	8	5	3	47.8	5.4	53.3	44.7	2.1	0.0	46.7
14	Type II	Kiseneke	853	0.58	74.7	0.90	14	9	5	35.7	16.3	52.0	26.4	0.4	21.2	48.0
17	Type II	Kiseneke	850	1.12	91.0	0.94	15	9	6	18.8	32.8	51.6	20.0	26.4	2.0	48.4
30	Type II	Mungi	700	0.59	61.7	0.87	13	10	3	30.7	20.6	51.3	17.8	30.9	0.0	48.7
21	Type II	Changa	830	1.84	47.4	0.84	11	6	5	37.9	13.3	51.2	10.6	4.4	33.8	48.8
33	Type II	Mungi	660	1.13	92.9	0.94	17	9	8	42.9	6.9	49.9	24.9	22.9	2.4	50.1
32	Type II	Mungi	704	0.79	83.3	0.90	16	7	9	31.2	16.8	48.0	9.1	4.1	38.8	52.0
20	Type II	Kiseneke	725	0.38	77.1	0.82	13	6	7	40.6	7.1	47.6	5.7	10.3	36.4	52.4
19	Type II	Kiseneke	735	0.18	52.1	0.70	9	5	4	43.4	3.1	46.5	6.7	3.9	42.9	53.5
16	Type II	Kiseneke	827	0.59	90.4	0.85	23	9	14	33.8	6.9	40.8	21.6	13.3	24.3	59.2
		Averadge	743	0.74	64.7	0.82	13.4	7.3	6.1	34.8	15.0	64.9	17.7	10.8	21.7	35.1
36	Type III	Ludewa	735	0.53	62.8	0.79	18	10	8	27.8	7.7	35.5	16.0	45.2	3.4	64.5
24	Type III	Mungi	782	0.52	67.9	0.69	11	8	3	16.3	16.1	32.4	0.0	2.5	65.1	67.6
10	Type III	Kiseneke	860	0.32	48.0	0.66	13	8	5	14.5	13.5	28.0	12.8	57.5	1.7	72.0
13	Type III	Kiseneke	830	0.95	89.9	0.98	24	11	13	19.3	8.5	27.8	35.5	14.2	22.5	72.2
34	Type III	Ludewa	701	0.79	89.8	0.79	9	4	5	24.0	1.7	25.8	35.4	17.1	21.7	74.2
40	Type III	Ludewa	649	0.70	48.1	0.58	9	5	4	24.2	0.8	24.9	15.5	59.6	0.0	75.1
15	Type III	Kiseneke	838	0.53	83.6	0.85	16	5	11	18.5	3.2	21.7	8.1	68.4	1.8	78.3
11	Type III	Kiseneke	860	0.34	61.8	0.72	12	5	7	3.0	13.9	16.9	14.0	33.2	35.9	83.1
		Averadge	782	0.58	69.0	0.76	14.0	7.0	7.0	18.8	7.0	26.6	19.6	42.2	12.4	73.4
8	Type IV	Kiseneke	902	0.43	75.6	0.84	16	6	10	5.9	8.7	14.6	16.4	35.1	33.9	85.4
2	Type IV	Kiseneke	1175	0.54	88.1	0.54	10	4	6	9.3	0.9	10.3	43.8	3.7	42.3	89.7

Table 1. *Cont.*

Household No.	Cropping Pattern	Village District	Altitude	Area	Percentage of cropping Area	Index of DIVERSITy *	Number of Crop Species			(3a) The Percentage of Area Occupied by Tree Crops			Tall Herbaceous Crop	Medium Herbaceous Crops	(3b) The Percentage of Area Occupied by Herbaceous Crops	
							Total	Tree Crops	Herbaceous Crops	Tall Tree	Short Tree	Total			Short Herbaceous Crops	Total
			(m)	(Acre)	(%)		(Spieces)	(Spieces)	(Spieces)	(%)	(%)	(%)	(%)	(%)	(%)	(%)
9	Type IV	Kiseneke	910	0.38	93.2	0.68	14	6	8	7.4	1.7	9.1	14.6	37.0	39.3	90.9
1	Type IV	Kiseneke	1222	0.70	80.3	0.73	7	2	5	4.4	2.3	6.7	21.3	41.7	30.2	93.3
12	Type IV	Kiseneke	852	1.23	37.7	0.65	13	6	7	4.6	1.7	6.3	33.2	29.9	30.6	93.7
7	Type IV	Kiseneke	920	0.53	88.8	0.58	10	2	8	2.6	3.7	6.3	15.0	38.7	40.1	93.7
4	Type IV	Kiseneke	1109	0.44	69.3	0.78	14	1	13	0.0	0.3	0.3	22.9	31.7	35.1	99.7
		Averadge	1013	0.6	76.1	0.69	12.0	3.9	8.1	4.6	2.2	7.7	24.7	32.1	36.4	92.3
25	Type V	Mungi	800	0.44	84.6	0.46	12	6	6	4.2	61.9	66.1	0.9	31.9	1.1	33.9
5	Type V	Kiseneke	950	0.21	65.6	0.48	6	2	4	7.3	0.5	7.8	20.8	63.3	8.1	92.2
6	Type V	Kiseneke	944	0.15	76.4	0.48	10	5	5	4.3	3.4	7.7	5.1	85.4	1.8	92.3
3	Type V	Kiseneke	1170	0.74	71.7	0.40	6	0	5	0.0	0.0	0.0	24.4	7.9	67.7	100.0
18	Type V	Kiseneke	736	0.17	82.9	0.29	3	0	3	0.0	0.0	0.0	7.5	13.9	78.6	100.0
		Average	905	0.54	76.2	0.42	11.5	4.6	6.8	7.9	26.3	16.3	3.0	30.8	32.0	83.7

* Shows the value of the exponent (D = (Σsin/N)²) of Simpson.

First, according to the plant's growth habit, we divided the crops planted in the homegardens as tree crops and herbaceous crops, as shown in Table 1. Then, we evaluated the ratios of the areas occupied by the respective crops. Next, the rates of cultivated area occupied by the crops (3a), and (3b) in each household's homegarden were plotted (Table 1). The ratios of (3a) and (3b) differed remarkably depending on the household (Table 1). Moreover, we divided the cropping patterns of tree crops (3a) into four types according to the area of the homegardens occupied by the crops: high (type-I, $\geq$70%), medium (type-II, 40–69%), low (type-III, 10–39%), and very low (type-IV, <10%).

3.2. Crop Grouping Based on Their Heights

Since the heights of tree crops differ remarkably depending on the variety and individual tree type, we divided the 23 varieties of tree crops into two groups: tall and short. The tall trees are 10 m or above and the short trees are less than 10 m (Table 2). Since herbaceous crops do not generally differ considerably in height by individual plant or variety, they were classified according to the average height of each crop. Herbaceous crops with a height of 5 m or more were classified as tall herbaceous crops (Table 3).

We report homegardens dominated by tree crops (type I and II). The ratios of the areas occupied by the five crop groups for each household classified according to height are shown in Tables 1–3. In cropping type I, tree crops covered 70% or more of the cultivated area within the homegarden; many households mainly planted tall trees such as coconut and jackfruit. An exception was the homegarden of household number 29, which primarily cultivated small trees, including clove and cinnamon. In cropping type II, tree crops covered 40% to 69% of the cultivated area; many households mostly cultivated tall trees, except household numbers 42 and 17, which cultivated more small trees. Homegardens classified as type III and IV were characterized by the dominance of herbaceous crops. Here, the households preferred more herbaceous crops than tree crops, and the heights of crops varied (Table 1). Banana crops (a tall herbaceous crop) dominated four homegardens (household numbers 13, 34, 2, and 12), whereas the other seven homegardens (household numbers 36, 40, 15, 11, 8, 9, and 1) presented a dominance of medium herbaceous crops such as maize, cassava, and taro. Common bean and rice plants dominated four homegardens (household numbers 2, 9, 7, and 4). Many households presented diversity index values higher than 0.7; others only reached values below 0.5. Therefore, we classified the homegardens with a diversity index below 0.5 as type V. Five households (numbers 25, 5, 6, 3, and 18) belonged to that category, with an average number of 7.4 crop species per household, remarkably smaller than the other households. In the homegardens of this type, which were characterized by a low level of diversity, we observed a shared tendency among households of cropping specific crops in a manner similar to a monoculture (Table 1). Household number 25 cultivated cinnamon; household numbers 5 and 6, cassava; and household number 18, pineapple (*Ananas comosus* (L.) Merr.) in the form of a monoculture.

3.3. Distribution of Households According to Cropping Patterns

The cropping pattern changes with the elevation of households. Many homegardens of type I located at low elevations, mostly cultivated tree crops (Table 1). The relationship between elevation and the quantity of each crop was investigated, and the results are shown in Figure 3a–d. Type IV households located at high elevations (>850 m) hardly cultivated any tree crops, except household number 18. Type I homegardens were located in Ludewa and Mungi village districts, whereas type IV homegardens occurred only in the Kiseneke village district. Type II and III homegardens were distributed at elevations within the interval between types I and IV.

Table 2. Classification of the 23 tree crops observed in the homegardens of the 43 households according to their height.

Tree Form-Plant Type	Species Number	English	Local Name	Species	Average Height (m)	Max (m)	Number Measured (Number)	Percentage Measured (%)	Total Observed (Number)
Tall trees More than 15 m	1	Coco palm	Minazi	*Cocos nucifera* L.	14.1	25.0	90	32.7	275
	2	Bread Fruit	Msherisheli	*Treculia africana* Decne.	13.6	19.0	64	59.3	108
	3	Jack Fruit	Fenesi	*Artocarpus heterophyllus* Lam.	11.8	20.0	75	66.4	113
	4	–	Mwembe ng'ong'o	*Sclerocarya birrea* (A. Rich.) Hochst.	14.6	19.4	16	88.9	18
	5	Eucalyptus	Maidini	*Eucalyptus* sp.	15.3	23.0	16	88.9	18
	6	East African mahogany	Mkanbazi	*Khaya anthotheca* (Welw.) C.DC.	13.2	25.0	22	44.9	49
	7	Kapok	Msufi	*Ceiba pentandra* (L.) Gaertn.	13.5	22.0	4	80.0	5
	8	Durian	Mduriani	*Durio zibethinus* L.	20.0	20.0	1	100.0	1
	9	Jambolan	Mzambarawe	*Syzygium cuminii* L.	16.0	16.0	1	25.0	4
Short trees Less than 10 m	11	African Oil Palm	Chikichi	*Elaeis guineensis* Jacq.	8.0	8.0	1	50.0	2
	12	Avocado	Mfukado	*Persea americana* Mill.	8.1	8.0	15	36.6	41
	13	Mango	Mihembe	*Mangifera indica* L.	5.0	23.0	2	100.0	2
	14	Clove	Amdarasini	*Syzygium aromaticum* (L.) Merrill & Perry	5.6	12.0	49	32.0	153
	15	Cinnamon	Amdarasini	*Cinnamomum zeylanicum* J.Presl	3.4	8.0	45	4.5	997
	16	Coffee	Buni	*Coffea arabica/robusta* L.	3.1	8.0	23	12.0	192
	17	Tree cassava	Chisamvu	*Manihot* sp.	1.8	2.0	3	25.0	12
	18	Bamboo	Mlanzi	*Bambusoideae* spp.	3.0	3.0	7	–	
	19	Orange	Mchenza	*Citrus reticulata* Blanco, 1837	2.9	8.0	14	70.0	20
	20	Papaya	Papai	*Carica papaya* L.	2.0	2.0	6	37.5	16
	21	Soursop	Stafeli	*Annona muricata* L.	4.0	5.0	3	75.0	4
	22	–	Mkaranga mti	*Bombax rhodognaphalon* A.Robyns	2.3	3.0	2	50.0	4
	23	–	Kitupa	*Tephrosia vogelii* Hook.f.	1.7	2.0	3	30.0	10

Table 3. Classification of the 24 herbaceous crops observed in the homegardens of the 43 households according to their height.

Form-Plant Type	Tree Type	Number of Species	English	Local Name	Species
Tall herbaceous crops	More than 5 m	1	Banana	Makowo	*Musa* spp.
		2	Black pepper	Pilipili mtama	*Piper nigrum* L.
		3	Oyster nuts	Mkweme	*Telfairia occidentalis* Hook.f.
		4	Cardamom	Iliki	*Elettaria cardamomun* (L.) Maton
		5	Coco yam	Ghimbi	*Colocasia esculenta* (L.) Schott/*Xanthosoma sagittifolium*
		6	Cassava	Gumuhogo	*Manihot esculenta* Crantz
		7	Castor oil plant	Mnyemba	*Ricinus communis* L
		8	Hyacinth bean	Fyifyi	*Lablab niger* (L.) Sweet
		9	Night shade	Diderega	*Basella alba* L.
Medium herbaceous crops	1–5 m	10	Maize	Ditama	*Zea mays* L.
		11	Okra	Dibamia	*Abelmoschus esculentus* (L.) Moench
		12	Passion fruits	Matunda kweme	*Passiflora edulis* Sims, 1818
		13	Pigeon pea	Zimange	*Cajanus cajan* L.
		14	Sisal	Mkonge	*Agave sisalana* Perrine
		15	Scarlet runner	Kikamba	*Phaseolus coccineus* L.
		16	Sugarcane	Mguwa	*Saccharum officinarum* L.
		17	Yam	Vigonzo	*Dioscorea alata* L.
		18	Amaranths	Gumchicha	*Amaranthus* spp.
		19	Basil	Not sure	*Ocimum basilicum* L.
		20	Bitter tomato	Zinungwi	*Solanum* spp.
		21	Chili	Pilipili	*Capsicum frutescens* L. *Capsicum chinense* Jacq. *Capsicum annuum* L.
Short herbaceous crops	Less than 1 m	22	Ginger	Mbwiga	*Zingiber officinale* Roscoe
		23	Kidney beans	Maharagi	*Phaseolus vulgaris* L.
		24	Pineapple	Dinanasi	*Ananas comosus* (L.) Merr.
		25	Rice	Uhunga	*Oryza sativa* L.
		26	Roselle	Damudamu	*Hibiscus subdariffa* L.
		27	Sweet potato	Tenbere	*Ipomoea batatas* (L.) Lam.
		28	Tomato	Nyanya	*Lycopersicon esculentum* L.
		29	Vanilla	Not sure	*Vanilla planifolia* Jacks. ex Andrews

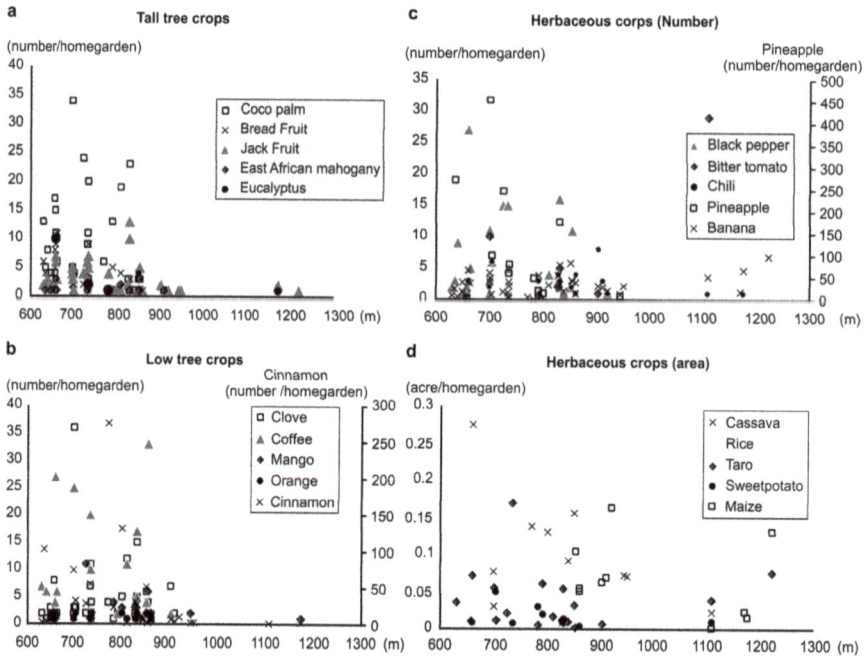

Figure 3. Relationship between altitude of homegardens and cultivated number or area occupied by each crop in each homegarden. (**a**) Tall tree crops. (**b**) Short tree crops. (**c**) Herbaceous crops (number). (**d**) Herbaceous crops (area).

3.4. Differences in Temperature at Each Elevation

The annual rainfall in the investigated village ranged between 1800–2000 mm [26]. On the eastern slopes, rainfall rarely limited crop growth even at comparatively low elevations; hence, differences in rainfall, but not in temperature, due to elevation strongly affected the variability of cultivated crops.

From September 2004 to January 2007, we registered average temperatures of 23.3, 22.2, and 21.1 °C at the elevations 650, 750, and 1000 m in Kiboguwa Village, respectively. At these three elevations, the temperature increased by 1.1 °C with decreasing elevation with no major differences observed across the years. Rice plant cultivation began in June and July when the lowest temperatures are recorded. The average lowest temperatures during this period were 15.5 °C at 1000 m; 15.8 °C at 750 m; and 17.3 °C at 650 m. Temperatures close to 15 °C damage rice plants, which we observed at an elevation of 1000 m. The average temperatures increase during the period from December to February, with the highest daily temperature of 33.1 °C, which is not considered high enough to potentially damage crops.

3.5. Characteristics of the Crops at Different Elevations

The temperature at low elevations is suitable for growing tree crops, except coffee. When we compared satellite images obtained in 2013 (Figure 4c) with those obtained in 2005 (Figure 4b), we did not observe any significant changes in the forest landscape at the site regarding its location, size, or condition. Additionally, we did not observe any major changes in the uses of land outside the homegardens. These findings suggested that, from 2005 onward, no climate changes occurred that could impact the cropping patterns within this region, and the phenomenon described in this study might continue to be observed even at present.

Considering the quantity cropped for each elevation (Figure 3a,b), although some households have nearly 70% of their cropping area occupied by cinnamon and coconut, the ratios of these cropping areas remarkably drop when the elevation approaches 800 m. Coconut is not cultivated at elevations of 900 m and above (Figure 3a). The cultivation of cinnamon extends to elevations of 1000 m and above (Figure 3b), but at a smaller proportion; the cropping quantity further decreases at low elevations. The same tendency occurred for clove, coffee, breadfruit, jackfruit, and East African mahogany (*Khaya anthotheca*) as shown in Figure 3a,b.

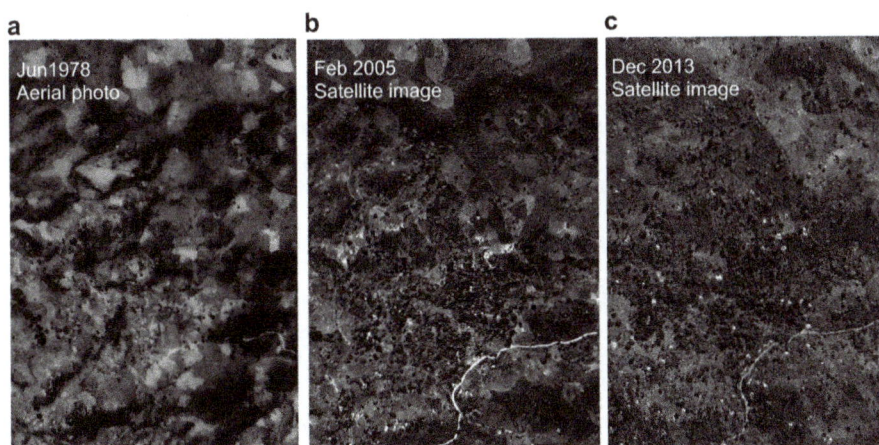

Figure 4. Effects of the Ujamaa policy on the development of forest-like landscape in the study village. Comparison of aerial photographs obtained in 1978 (National Information purchased from Office for Dar es Salaam) and satellite images obtained in 2005 and 2013 (purchased from Pasco Corporation; Quick bird pan-sharpened image). The dotted boxed portion indicates the satellite images and the aerial extent of the image shown in Figure 4.

Regarding the cultivation of herbaceous crops at different elevations, we observed the cultivation of maize limited to homegardens located at elevations equal to or higher than 850 m (Figure 3d). Common bean is cropped as a monoculture in sloping fields, and people from Kiboguwa often planted it at elevations of 900 m to 1000 m. Legumes cultivated in this village are better adapted to the environment [27]. Maize and common bean grow together in the homegardens distributed at elevations of 800 m and above. Conversely, cassava and rice plants are cropped at an elevation belt different from that of maize and common bean (Figure 3d). They are frequently cultivated together on sloping fields.

Since the cultivation characteristics of maize, rice, and cassava differ remarkably depending on the variety of each crop, the order of elevation belts in which they are cultivated also relies on the combination of varieties selected. However, in the mountain regions where both maize and rice plants can be cropped together, rice plants are generally cultivated at a lower elevation belt.

3.6. Variations in the Cropping Patterns of Households Distributed at Elevations of 900 m and Below

In addition to the differences in cropping patterns according to elevation, major differences existed in homegarden cropping patterns between adjacent households distributed at low elevations of 600–900 m. Hence, temperature differences due to elevation alone did not explain these differences. The households from number 10 onward are located at elevations of 900 m and below (Figure 1b).

3.7. Other Cropping Patterns

Some households such as number 34 and 35 presented a unique cropping pattern that did not depend on elevation. Household numbers 34 and 35 are located in the Ludewa village district at low elevations of 701 m and 658 m with homegarden areas of 0.79 and 0.37 acres, respectively. Unlike others, household 34 is dominated by rice, with many banana trees in front of the house (Figure 3a). Moreover, distributed within the rice fields, we observed large trees such as coconut (>20 m), clove, and jackfruit (about 10 m). Conversely, household number 35 (Figure 3b), located 50 m to the north of household number 34, possessed many tall tree crops (>20 m) and breadfruit trees within a small plot that forms a forest landscape.

3.8. Historical Changes of the Homegarden Crops: Introduction and Spread of Tree Crops

The results of the interviews conducted suggested that, before the introduction of spices, they were brought by an Arab who had settled in a town called Matombo, approximately 30 km from Kiboguwa. However, no descriptions were found to support this. The growing of various tree crops gradually and recently introduced in homegardens spread to create a forest landscape at low elevations in this region.

Around a century ago, practically no tree crops currently found in Kiboguwa existed. Before the German colonial period, the only perennial crops in this region were mango and banana. The slave trade by Arabs at Zanzibar Island under the rule of Omani occurred in both directions on the north and south sides of the Uluguru Mountains [28], which led to the belief that Arabs brought mangoes to this region. Even in an aerial photograph obtained in 1964, large mango trees can be found on the sloping fields.

3.9. Effect of Ujaama on the Expansion of Tree Crops in Homegardens

We compared satellite photographs immediately after the Ujaama policy introduction (Figure 4a), and 30 years later (Figure 4b) to see the forest landscape changes in the study area. We found that Mungi and Changa village districts developed a forest landscape within the surveyed Kuboguwa village. Such a forest landscape did not exist in the 1978 photograph when the forced migration began, caused by the Ujamaa policy (Figures 4a and 5). The photograph from 1978 confirmed the growth of tree crops around the homegardens in the settlements of Ludewa, but, with a lower number of trees than those found at present. This suggests that, in this period, most of the forest landscape developed in the village, present in more recent pictures, was very small and did not appear in the aerial view. There were no tree crops, except banana, in the region until around five decades ago when the tree crops were introduced from Zanzibar Island. Therefore, 50 years ago, the landscape of homegardens differed completely from that indicated at present.

Figure 5 shows plots of the households in the four village districts and how they obtained their homegardens. The results reveal households which were forced to migrate, received their homegarden parcel from the village government, and continue to live on that land at present. Even within the same village, forced migration through the Ujamma policy affected some districts. Some of the migrated people stayed in the lower part, and some of them returned to their original lands in the upper part. Although some villagers were forced to relocate to the lower part of the district, they kept their original property in the upper part of the Kiseneke district. Consequently, the Ujamaa policy hardly affected the Kiseneke village district (Figure 5). Conversely, in the other three districts—Ludewa, Mungi, and Changa—many people who lived along the ridge running through the center of the village were forced to migrate to locations with no previous settlement and continued to live in these places. Therefore, these village districts became places with a better-developed forest landscape and the Ujamaa policy played a vital role in this regard.

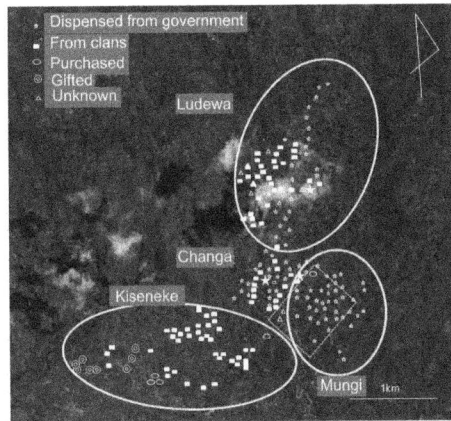

Figure 5. Distributions of 258 households in the four village districts shown using different symbols according to the methods of how the lands were obtained. The symbol (☆) indicates the households which were forced to migrate, received their homegarden plot from the village government, and continue to live on that land. Quick bird image (2005).

3.10. Effects of the Ujamaa Policy on the Land Outside the Homegardens

According to the survey results of 80 households in 2007, 37 of 80 households (46.3%) obtained their homegarden land through inheritance, either paternal or maternal. The next most frequent source was distribution from the government (32 households, 40.0%). The Socialist policy was implemented in Tanzania from the 1960s to the first half of the 1970s. As part of that policy, in the construction and promotion of Ujjama village, people living in a dispersed way increased agricultural productivity (Hyden, 1980). From 1971 onwards, a semi-compulsive migration strategy (Operation Vijijini) was initiated nationwide. In Kibogwa village, in 1974, the land of a specific clan (exactly Lineage) was confiscated by the Tanzanian government and the residence was redistributed including households not belonging to the clan. In 2007, 32 households, about 40% of the surveyed 80 households, lived on land granted by or inherited from the government. Most of the clan's land that was confiscated by the government is residential, and most of the cultivated land preserved the customary land holding, as established from the interviews with the elderly. In addition, it seems that the collection was carried out in units of village districts; the residence seems to have not moved beyond the scope of the village; the original villagers use the cultivated land of the village to settle. This did not change before or after.

Many households divide these fields into parcels and cultivate three parcels per year. Many households cultivated rice in the fields on the slopes of the ridge away from the high concentration of homegardens. According to the survey results of 80 households in 2007, households cultivated 268 land parcels. Among them, 197 parcels (73.5%) were obtained via the clan by the present cultivators. The 44 parcels (16.4%) were cultivated in different ways: 26 parcels were leased for cash, 17 were purchased, and only 1 parcel was a sloping field allocated by the government. Information on how the remaining 27 parcels were obtained was unknown. Accordingly, the Ujamaa policy does not have any practical effect on how sloping fields are used at present. Of the 268 parcels, 65 (24.25%) contained maize; 94 (35.1%), rice; and 78 (29.1%), cassava, together representing 88.45% of the total cropping. Sloping fields used as fallow or for the cultivation of other crops constituted only three parcels (1.12%), but they did not cultivate perennial crops. The region's matrilineal clan plays a substantial role in the distribution of sloping fields and their methods of use. Consequently, the cultivation of perennial crops such as tree crops and banana remains limited in the sloping areas.

Following the land tenure system in the study area, it is thought to be difficult to plant perennial crops such as banana or tree crops in the fields on the slopes. Therefore, on sloping fields that extend outside the homegardens, hardly any tree crops are planted [20]. The sloping fields have been used for the cultivation of short cycle crops of staple food such as maize, rice, or cassava, under the influence of traditional land tenure systems possessed by Matrilineal clans [20]. From these results, it was determined that on the lands which were distributed by the village government, the influence of the traditional land tenure system was removed and people started to cultivate perennial crops on their own land.

However, some homegardens might cultivate just one of the staple food crops, as in the case of household number 34, which planted rice crops in the entire homegarden. Maize is cultivated at a comparatively high elevation within the village; in many cases, it is cultivated at an elevation of 800 m or above on the large ridge in the Kiseneke village district that divides the village in half. Rice plants and cassava are frequently cultivated at elevations of 800 m and below. In particular, rice plants are often cultivated on the sloping fields to the south of the area where the homegardens are concentrated in the Mungi village district (Figure 5). On such sloping fields, the land-use pattern for the cultivation of staple food crops reveals similarities between the households in this village: fixed location and growing season. In addition to the land tenure system centered on matrilineal clans, there is a cropping system based on the forms of regional meals that also defines the crop varieties planted in the sloping fields. Therefore, the cropping patterns in homegardens differ from the sloping fields.

3.11. Changes of the Tree Crop Cultivation

By comparing the satellite images of December 2013 with those in 2005, we traced the direction of the Kiboguwa Village forest landscape development (Figure 4b,c). We identified more abundant vegetation with newly constructed houses in the 2013 image. However, many of the houses visible in the image of 2005 still exist after eight years, and the forest landscape also covered practically the same range (Figure 4b,c). In 2005 and 2013, the upper region in the image mainly showed the cultivation of rice with hardly any trees occurring in this place. In Kiboguwa Village, some of the homegardens are abandoned due to the death or migration of the head of the household. In Kiseneke, there was a relatively frequent migration tendency, and the places used as homegardens frequently became unmanned and fell into an abandoned state. These aforementioned abandoned states are known as Biamo.

Of the 55 households in Kiseneke surveyed in 2005, five established homegardens at separate locations or moved to a different household to start living at this new location in 2006. Of these five households, in two households, the wife returned to her original home due to divorce. In another household, an elderly lady lived alone and, to live with her son, she established a homegarden in a more central part of the village and moved there. Therefore, the location of a previous homegarden returned to its original characteristics. In the remaining two households, one household had poor relations with the neighbors and moved to another village and, in the other household, a sister with a deteriorated relationship with her brother established a new home in the field of her husband. Therefore, although the locations of the homegardens changed, the land use in the village remained likely the same over the eight years.

3.12. Characteristics of the Household Structures on Cropping Patterns

The households with homegardens distributed at elevations of 900 m and below are arranged according to the ages of their oldest member (Table S1). In the study area, households presented a strong tendency to have a nuclear family. Household numbers 14, 35, and 40 were occupied by widows, their mothers, and children (Table S1).

Except for household numbers 14, 35, and 40, the cropping patterns shift following each change of generation. Married couples established new homegardens in fields inherited from either the parents of the wife or the husband. However, because only annual crops are planted in the inherited fields, the couples plant new crops according to their preference. Therefore, the cropping pattern in homegardens reflected the preferences and necessities of the married couples. We considered this as one of the leading factors to the generation of diversity in the homegarden cropping patterns among households.

When evaluated the homegardens according to the age of the eldest member, who we considered as the head of the household; we observed that the method of obtaining homegardens differed by age group. The heads of eight households were in their seventies or older. Six households obtained their homegarden during the forced migration in 1974, in their 60s. The head of household number 35 was in her 50s or younger, and the heads of household numbers 24 and 28, answered that they migrated to the land allocated to them by the government.

Under the Ujamaa policy, immigrants supposedly received fixed and equally distributed areas of land. However, the head of household number 20 obtained a narrower homegarden in comparison to those of the other households. The head of this household had four wives, of whom three lived at an adjacent homegarden. Furthermore, his sons lived together with these three wives at household numbers 18 and 19 and had established homegardens close to that of their father. All of these lands formerly belonged to the head of household number 20.

In summary, age group explains how homegardens differed in terms of how they were acquired and their size. This suggested that the older age group tended to receive a bigger homegarden. The ratio of the cropping area for tree crops tended to be higher among the older age groups.

4. Discussion

We identified the cropping patterns of the homegardens in Kiboguwa village, located in the mountainous area of central Tanzania, through a survey of participant observations accompanied by a long-term stay. This allowed us to evaluate the critical factors related to the cropping patterns such as geography, climate, history, land use policies, and the cultural behavior on land use, particularly the Ujamaa policy.

The crops cultivated in the homegardens in the research area changed over time. Previous reports suggest that Bananas were introduced in the distant past, and an English explorer from the 19th century described how the Luguru people wore the leaves of banana plants instead of clothes [21]. Coffee also arrived there a long time ago, with records of a German who began cultivating coffee in the 1890s on the western slopes of Mgeta [29]. However, the route by which coffee was introduced into the village studied (Kiboguwa) remains unknown. A remarkable increase in the number of tree crops in this region likely occurred around the time of Tanzania's independence, 60 years ago. In the same period, spice crops such as cinnamon and clove, which are cultivated as staple food crops, were also introduced into this region. This introduction likely resulted from the poll tax imposed by British colonial rule. Many people wanted to escape the poll tax and, at that time, the country was still divided and conquered by the Arabs and British [30]. Hence, people headed for the Arab Zanzibar Island which was not within the scope of the poll tax. On Zanzibar Island, they became seasonal workers on the clove plantations established by the Arabs, and the Luguru people visited the island together with the Sukuma and Nyamwezi people who lived in the northern part of Tanzania [28]. Before independence, the people who went to Zanzibar Island subsequently returned to the region studied in this research and brought with them many crops, which might explain the sudden increase in the number of trees crops. When coconuts were introduced also remains unclear. Since slaves on Zanzibar Island used coconut plantations [28], the introduction of coconut might have occurred similarly as that of clove; they arrived in this region around the time of independence.

At present, many homegardens have created an expansive forest landscape centered on the Changa, Mungi, and Ludewa village districts that developed on land excluded from the allocation of land to clans during the implementation of the Ujamaa policy. Following the land tenure system in the study area, it is thought to be difficult to plant perennial crops such as banana or tree crops in the fields on the slopes. Therefore, on sloping fields that extend outside the homegardens, hardly any tree crops are planted there [20]. Land distributed by the village government under the Ujamaa policy removes the influence of traditional land tenure, and people started to cultivate perennial crops including banana in their homegardens. About 60 years ago, at the time of the immigration under the Ujamaa policy, the cropping pattern of the homegardens, less than 900m above sea level, suitable for many tree crops, became diverse.

In addition, ecological differences which are observed at different altitudes of the slopes increase the diversity of the cropping patterns of the homegardens in the village. Commercial crops such as cinnamon, clove, coffee, and coconut were the most cultivated, many of which require a damp environment with a high temperature for growth. The suitable conditions for growing cinnamon are an average temperature of between 20 °C and 30 °C; annual rainfall of 1250–2500 mm, and an elevation of 300–350 m. The suitable conditions for growing clove are a temperature between 24 °C and 33 °C with annual rainfall of 2000–3000 mm, and an elevation range of 300–600 m [31]. Moreover, the suitable conditions for growing coconut are an average temperature of 27 °C to 28 °C, but not below 20 °C, similar to cinnamon. The monthly rainfall should be an average of 130 mm (around 50 mm in the dry season); therefore, in Eastern Africa, these crops can be grown at elevations of 1100 m and below (Weis, E.A. 2002). Many of the tree crops cultivated in this village were suitable for cultivation at low altitude in the village. The cultivation of herbaceous crops was also observed at altitudes suitable for their cultivation characteristics.

Furthermore, the family structure and custom homegarden land use of the village also contributed to the diversity of the homegarden cropping patterns. Households in this village are often composed of nuclear families, with one homegarden per household. Typically, after abandonment, a homegarden becomes indistinguishable from a normal field. However, the cropping patterns of an abandoned homegarden for only one generation still reflected the ideas of the people living there, their ages, and household structure. This likely increased the diversity of homegarden cropping patterns within the study area. In some regions of the islands of Southeast Asia, people manage crops in their homegardens over many generations. In these regions, homegardens become places where culture flows from one generation to another [32]. The cropping patterns noted in the Southeast Asia homegardens, which mainly grow banana and coffee, differed from those indicated in our study: their cropping patterns are less likely to reflect factors such as the ages, preferences, and health conditions of the family members. Additionally, the tendency to pass down homegarden cropping patterns through generations can be noted in the Kilimanjaro and Usambara Mountains [33].

5. Conclusions

The diversity of the cropping patterns observed in the homegardens in this study village was influenced by factors related to regional characteristics such as the regional history and the customs and policies. The introduction of tree crops began with people returning from Zanzibar island, before and after the colonial era, who brought back various crops to this area. Furthermore, the customs concerning the ownership of lands possessed by maternity clans and the ownership of trees has been partly broken by the Ujama policy and tree crop cultivations started to spread in homegardens in the village. In addition, households are often composed of nuclear families, and the land of homegardens tends to be abandoned by one generation. Such factors contribute to the diversity of cropping patterns. In addition, ecological diversity distributed on the slopes of the Mountains from around 650 m to around 1200 m, also makes the cropping pattern diverse. In homegardens distributed above 900 m, only a few tree crops were observed. Instead of tree crops, herbaceous crops such as maize or common bean or banana were cultivated. Because bananas are not a main staple food crop for the people of Kiboguwa village, there are some homegardens in which only few bananas were cultivated. This point is not the same as the homegardens of people in the northern part of Tanzania who are called banana eaters. Therefore, it can be said that cultural differences in terms of staple food crop affected the diversity of cropping patterns in both areas. The contents or methods of agricultural and rural development projects or the purpose of research on tropical agriculture must be relevant to actual situations [34]. In agricultural technological supports, the agriculture of the target area of the projects is analyzed mainly from economical or agro-ecological point of view [35]. Factors influencing cropping patterns are assumed to be different in the region, and the contents of agricultural technical supports targeting the hormegardens must be different among these regions. Without knowing suitable and specific points of view for each area, the nature or real background of the phenomenon could not be understood. Analysis using an inductive method—considering the background against which the phenomenon becomes evident after collecting the information from the target area in this manner—is thought to lead to an essential understanding. By investigating the cropping patterns of homegardens, we provided information to support future developments that improve the agricultural development in such rich Tanzanian homegardens. In this study, we tried to understand the relationship between the phenomenon appearing as cropping patterns from the viewpoint of the investigator and the factors acting on it. However, information on how farmers' own thinking and intention about cropping of the homegardens have not been collected. Such research is also considered to be necessary.

Supplementary Materials: The following are available online at http://www.mdpi.com/2077-0472/8/9/141/s1, Table S1: Information on household members who have lived together at a homegarden and characteristics of cropping patterns of each homegarden.

Author Contributions: J.K. and Y.Y. conceived and designed the experiments; K.I. performed the experiments; and Y.Y. analyzed the data; Y.Y. contributed reagents/materials/analysis tools; Y.Y. wrote the paper.

Funding: This study was funded by the Ministry of Education, Culture, Sports, Science and Technology Grant-in-aid for Scientific Research (grant number: 16101009; Representative: Kakeya Makoto) "Comprehensive research on African-type rural development based on regional research". The findings reported in this study are part of this research.

Conflicts of Interest: The authors declare no conflict of interest.

References

1. Nair, P.K.R.; Kumar, B.M. Introduction. In *Tropical Homegardens: A Time-Tested Example of Sustainable Agroforestry*; Kumar, B.M., Nair, P.K.R., Eds.; Springer: Dordrecht, The Netherlands, 2006; pp. 1–12.
2. Peyre, A.; Guidal, A.; Wiersum, K.F.; Bongers, F. Dynamics of homegarden structure and functions in Kerala, India. *Agrofor. Syst.* **2006**, *66*, 101–115. [CrossRef]
3. Wiersum, K.F. Diversity and change in homegarden cultivation in Indonesia. In *Tropical Homegardens: A Time-Tested Example of Sustainable Agroforestry*; Kumar, B.M., Nair, P.K.R., Eds.; Springer Science: Dordrecht, The Netherlands, 2006; Volume 3.
4. Galhena, D.H.; Freed, R.; Maredia, K.M. Home gardens: A promising approach to enhance household food security and wellbeing. *Agric. Food Secur.* **2013**, *2*, 8. [CrossRef]
5. Abebe, T.; Wiersum, K.F.; Bongers, F. Spatial and temporal variation in crop diversity in agroforestry homegardens of southern Ethiopia. *Agrofor. Syst.* **2010**, *78*, 309–322. [CrossRef]
6. Soda, O. *The History of Modern Agricultural Thought—For Agriculture in the 21st Century*; Iwanami Shyoten: Tokyo, Japan, 2013; p. 236. (In Japanese)
7. Suzuki, S. *Agricultural Extension in Developing Countries and Its Evaluation*; Tokyo Agricultural University Publishing Association: Tokyo, Japan, 2010; pp. 230–247. (In Japanese)
8. Tsuruta, T. Historical consideration of agriculture/pastoral complexes in Western African semi arid areas; case examples of the Tanzania ground strip and its surroundings. *Bull. Fac. Agric. Kinki Univ.* **2011**, *44*, 97–114. (In Japanese)
9. Zimpita, T.; Biggs, C.; Faber, M. Gardening Practices in a Rural Village in South Africa 10 Years after Completion of a Home Garden Project. *Food Nutr. Bull.* **2015**, *36*, 33–42. [CrossRef] [PubMed]
10. Assefa, T.T.; Jha, M.K.; Reyes, M.R.; Schimmel, K.; Tilahun, S.A. Commercial Home Gardens under Conservation Agriculture and Drip Irrigation for Small Holder Farming in sub Saharan Africa. In Proceedings of the ASABE Annual International Meeting, Spokane, WA, USA, 16–19 July 2017.
11. Kim, D.G.; Terefe, B.; Girma, S.; Kedir, H.; Morkie, N.; Woldie, T.M. Conversion of home garden agroforestry to crop fields reduced soil carbon and nitrogen stocks in Southern Ethiopia. *Agrofor. Syst.* **2017**, *90*, 251–264.
12. Gbedomon, R.C.; Salako, V.K.; Fandohan, A.B.; Idohou, A.F.R.; Kakaï, R.G.; Assogbadjo, A.E. Functional diversity of home gardens and their agrobiodiversity conservation benefits in Benin, West Africa. *J. Ethnobiol. Ethnomed.* **2017**, *13*, 66. [CrossRef] [PubMed]

13. Amberber, M.; Argaw, M.; Asfaw, Z. The role of homegardens for in situ conservation of plant biodiversity in Holeta Town, Oromia National Regional State, Ethiopia. *Int. J. Biodivers. Conserv.* **2014**, *6*, 8–16.

14. Abebe, T. Diversity in Homegarden Agroforestry Systems of Southern Ethiopia. Ph.D. Thesis, Wageningen University, Wageningen, The Netherlands, 2005.

15. Maruo, S. People who live with bananas: A case study of the village of the Haya in Western Tanzania. In *African Farmer's World—Transformation and Its Indigenous*; Kakeya, M., Ed.; Kyoto University Press: Kyoto, Japan, 2002; pp. 51–90. (In Japanese)

16. Reyes, T.; Quiroz, R.; Msikula, S. Socio-economic comparison between traditional and improved cultivation methods in Agroforestry systems, East Usambara Mountains, Tanzania. *Environ. Manag.* **2005**, *36*, 682–690. [CrossRef] [PubMed]

17. Rugalema, G.H.; Okting'Ati, A.; Johnsen, F.H. The homegarden agroforestry system of Bukoba district, North-Western Tanzania. 1. Farming system analysis. *Agrofor. Syst.* **1994**, *26*, 53–64. [CrossRef]

18. Soini, E. Land use change patterns and livelihood dynamics on the slopes of Mt. Kilimanjaro, Tanzania. *Agrofor. Syst.* **2005**, *85*, 306–323. [CrossRef]

19. Masawe, J.L. Farming system and agricultural production among small farmers in the Uluguru Mountain area, Morogoro region, Tanzania. *Afr. Study Monogr.* **1992**, *13*, 171–183.

20. Yamane, Y.; Kularatne, J.; Ito, K. Agricultural production and food consumption of mountain farmers in Tanzania: A case study of Kiboguwa village in Uruguru Mountains. *Agric. Food Secur.* **2018**, *7*, 54. [CrossRef]

21. Young, R.; Fosbrooke, H. *Land and Politics among the Luguru of Tanganyika*; Routledge & Kegan Paul: London, UK, 1960.

22. Beidelman, T.O. *The Matrilineal Peoples of Eastern Tanzania*; International African Institute: London, UK, 1967.

23. Yamane, Y.; Higuchi, H. Subsistence strategy of small farmers observed in planting of commercial crops in the homestead of the mountains of rural Tanzania: A case study of Kiboguwa Village on the Eastern Slopes of Uluguru Mountain in Tanzania. *J. Agric. Dev. Stud.* **2016**, *27*, 1–12. (In Japanese)

24. Maxwell, J. *Qualitative Research Design: An Interactive Approach*; Applied Social Research Methods; SAGE Publications: London, UK, 2005; pp. 87–121.

25. Krebs, C.J. *Ecological Methodology*, 2nd ed.; Group Limited Wesley, Longman: Harlow, UK, 1998.

26. Rwezimula, F.; Tanaka, U.; Ikeno, J. Agro-ecological characteristics and inherent roles of indigenous farming systems on the eastern slopes of northern Uluguru Mountains, Tanzania. *J. Agric. Dev. Stud.* **2010**, *20*, 65–71.

27. Arcland, J.D. *East African Crops*; Longman Group Limited: Hong Kong, China, 1980.

28. Odhiambo, E.S.A.; Ouso, T.I.; Williams, J.F.M. *The Inter Years: Tanganika, Zanzibar and Rwanda-Urundi Pages: A History of East Africa*; Longman Group Limited: Guangdong, China, 2006.

29. Ponte, S. Trapped in decline? Reassessing agrarian change and economic diversification on the Uluguru Mountains, Tanzania. *J. Mod. Afr. Stud.* **2001**, *39*, 81–100. [CrossRef]

30. Pretty, J. *Agri-Culture-Reconnecting, People, Lands and Nature*; Earthscan: London, UK, 2002.

31. Ravindran, P.N.; Nirmal-Babu, K.; Shylaja, M. Cinnamon and Cassia 'The Genus Cinnamomum'. In *Medicinal and Aromatic Plants Industrial Profiles*; CRC Press: Boca Raton, FL, USA, 2004.

32. Galluzzi, G.; Eyzaguirre, P.; Negri, V. Home gardens: Neglected hotspots of agro-biodiversity and cultural diversity. *Biodivers. Conserv.* **2010**, *19*, 3635–3654. [CrossRef]

33. Rao, M.R.; Rao, B.R.R. Medical plants in tropical homegardens. In *Tropical Homegardens: A Time-Tested Example of Sustainable Agroforestry*; Kumar, B.M., Nair, P.K.R., Eds.; Springer: Dordrecht, The Netherlands, 2006; pp. 205–232.

34. Angelsen, A.; Kaimowitz, D. Rethinking the causes of deforestation: Lessons from economic models. *World Bank Res Obs.* **1999**, *14*, 73–98. [CrossRef] [PubMed]

35. Nishimura, Y. Agriculture Rural development and technology development Technology transfer. In *Introduction to International Development Studies: Interdisciplinary Construction of Development Studies*; Otubo, S., Kimura, H., Ito, S., Eds.; Kusakasybo Limited: Saitama, Japan, 2009; pp. 334–343.

MDPI

St. Alban-Anlage 66

4052 Basel

Switzerland

Tel. +41 61 683 77 34

Fax +41 61 302 89 18

www.mdpi.com

Agriculture Editorial Office

E-mail: agriculture@mdpi.com

www.mdpi.com/journal/agriculture

www.ingramcontent.com/pod-product-compliance
Lightning Source LLC
Chambersburg PA
CBHW051900210326
41597CB00033B/5962